Roger H. Stuewer

copy

UMN Historian
of physics

STUEW

RUDOLF PEIERLS AND THEORETICAL PHYSICS

The editors and publishers gratefully acknowledge the permission of the artist, Mark Boxer, and the Warden and Fellows of New College, Oxford to reproduce the drawing of Sir Rudolf Peierls, on the front cover.

RUDOLF PEIERLS AND THEORETICAL PHYSICS

PROCEEDINGS OF THE SYMPOSIUM HELD IN OXFORD, ON JULY 11TH & 12TH 1974, TO MARK THE OCCASION OF THE RETIREMENT OF PROFESSOR SIR RUDOLF E. PEIERLS, F.R.S., C.B.E.

Edited by

I.J.R. AITCHISON and J.E. PATON

Department of Theoretical Physics, University of Oxford

PERGAMON PRESS

OXFORD · NEW YORK · TORONTO · SYDNEY · PARIS · FRANKFURT

U.K. Pergamon Press Ltd., Headington Hill Hall, Oxford OX3 0BW, England

U.S.A. Pergamon Press Inc., Maxwell House, Fairview Park, Elmsford, New York 10523, U.S.A.

CANADA Pergamon of Canada Ltd., 75 The East Mall, Toronto. Canada

AUSTRALIA Pergamon Press (Aust.) Pty. Ltd., 19a Boundary Street, Rushcutters Bay, N.S.W. 2011, Australia

FRANCE Pergamon Press SARL, 24 rue des Ecoles, 75240 Paris, Cedex 05, France

WEST GERMANY Pergamon Press GmbH, 6242 Kronberg-Taunus, Pferdstrasse 1, Frankfurt-am-Main, West Germany

First edition 1977

Library of Congress Cataloging in Publication Data

Peierls Symposium, Oxford, 1974.
The Peierls Symposium.

1. Physics—Congresses. 2. Peierls, Rudolf Ernst, Sir, 1907– I. Peierls, Rudolf Ernst, Sir 1907–
II. Aitchison, Ian Johnston Rhind, 1936–
III. Paton, J. E.
QCI.P3 1974 539 75-41493
ISBN 0-08-020606-9 Hardcover
0-08-021621-8 Flexicover

In order to make this volume available as economically and rapidly as possible the author's typescript has been reproduced in its original form. This method unfortunately has its typographical limitations but it is hoped that they in no way distract the reader.

Printed in Great Britian by A. Wheaton & Co. Exeter

CONTENTS

PREFACE

The Peierls Symposium, in honour of Sir Rudolf Ernst Peierls on the occasion of his retirement from the Wykeham Chair of Theoretical Physics in the University of Oxford, was held at Oxford on July 11th & 12th, 1974. The two hundred participants all had some professional connection with Peierls, and the invited speakers were each asked to talk on some development in physics with which Peierls had been particularly associated. It is these invited talks, together with some of the discussion that followed them, which are reproduced in this volume. They range in style from the historical reminiscence to the up-to-date review survey. The topics covered span a wide range of theoretical physics, including nuclear theory, statistical mechanics, solid state physics and elementary particle physics, mirroring the great breadth of Peierls' own interests.

The invited talks were recorded, and from the tapes we produced a draft version, which was put into final form by the speakers themselves. We are greatly indebted to them for the care and speed with which they corrected the draft manuscripts.

We deliberately asked the speakers not to eliminate all colloquialisms. The occasion had the spirit, which we wished to preserve, of a gathering of an "extended family", the creation of which may be regarded as a significant part of Peierls' scientific achievements. We hope that the personal flavour of the contributions will still come across, even in this printed form.

We know we speak for very many physicists, and their families, all over the world, when we offer to Professor Sir Rudolf and Lady Peierls our sincere thanks and very best wishes.

Oxford, May 1976

I. J. R. Aitchison
J. E. Paton

1

The Scattering of Pions by Nuclei

H. A. BETHE

I want to talk about the scattering of pions by nuclear matter and by finite nuclei. It is assumed in this that you know the scattering by a nucleon, and the question is to derive the scattering by a complex system from the scattering by one nucleon.

We begin with an equation which is familiar from the theory of the refractive index n in electrodynamics and optics, namely that the wave number in the medium k is related to the wave number in vacuum k_o by the equation

$$k^2 = k_o{}^2 + 4\pi\rho f(k, 0) \tag{1.1}$$

which contains as the main part the forward scattering amplitude f(k,o). The k_o, of course, in the case of pions, is related to the energy ω by the equation

$$k_o{}^2 = \omega^2 - \mu^2 \tag{1.2}$$

If I consider pions of energy, let's say, up to 300-400 MeV, which is what I am interested in here, then the scattering is mostly by S- and P-waves and therefore the forward scattering amplitude is expected to be of the form

$$f(k,o) = a_o + a_1 k^2 \tag{1.3}$$

where the constant a_1 represents the P-wave scattering.

Now if you put these things together, and this has been done many years ago - although not in exactly this form - then you find k^2 in terms of the energy

$$k^2 = \frac{\omega^2 - \mu^2 + 4\pi\rho a_o(\omega)}{1 - 4\pi\rho a_1(\omega)} \qquad (1.4)$$

and the important part here is the denominator which is $1 - 4\pi\rho a_1(\omega)$ where ρ is the density of scatterers (density of nucleons) and $a_1(\omega)$ is the P-wave scattering amplitude (apart from the factor k^2). This is a minor catastrophe; namely, a_1 might very well be just big enough so that the denominator becomes zero. You may say it can't become zero because a_1 is complex - the scattering amplitude is complex - but nevertheless we are still in trouble. For one thing, at very low energy the scattering amplitude is real; but for another thing, the imaginary part of the scattering amplitude always tells you the probability of having processes which remove the particles - the pions - from the incident beam. Now in the case of pion scattering by an assembly of nucleons, that is the probability of scattering in a collision with a single nucleon which is usually called a quasi-elastic collision if it happens in a nucleus. Now the imaginary part of $a_1(\omega)$ therefore surely is different when the nucleons are inside the nucleus from what it is for free nucleons, because for free nucleons I have no impediment to scattering and inside the nucleus I have. So it is surely wrong to consider a_1 to be the same quantity which you derive from the scattering of a pion by a free nucleon. Now if I take zero energy then the denominator is indeed zero, at a density of $\frac{1}{3}\mu^3$, which is just about two-thirds of the density of normal nuclear matter. The density of normal nuclear matter is .16 per $(\text{Fermi})^3$ and that's just about one-half μ^3. (μ is the reciprocal Compton wavelength of the pion and is about .7 inverse Fermis).

So this has been the status of the subject, more or less, for some twelve years, until a paper by Barshay, Rostokin and Vagradov (1) in 1973. Another paper putting the same idea in a much simpler and more transparent fashion followed, by Barshay, Brown and Rho (2). They took the point of view that really what happens in pion-nucleon scattering is essentially that the pion makes the nucleon into a resonant particle, into a Δ particle, and they then proved that this is the only part of the pion scattering which will survive inside nuclear matter. Then you automatically find that your $a_1(\omega)$ contains the wavenumber of the pion inside nuclear matter and everything is all right. What I have to say today is rather in the spirit of this but it goes more explicitly into the pion dynamics.

The fundamental theory of the pion scattering in nuclear matter was developed by Dover and Lemmer, published about a year ago in Physical

Review (3). They made the very simple statement that what you have to calculate is the self-energy of the pion (which is given by the diagram in Fig. 1.1(a)) inside the nuclear matter, and you then have in this diagram a loop of a nucleon, which you can cut at one or two points. If you cut it at one point you get the diagram Fig. 1.1(b), which is the familiar scattering diagram for pions which is responsible for (3,3) scattering, and if you cut it at the other point, you get the diagram of Fig. 1.1(c), in which you get first absorption, and then re-remission of the pion.

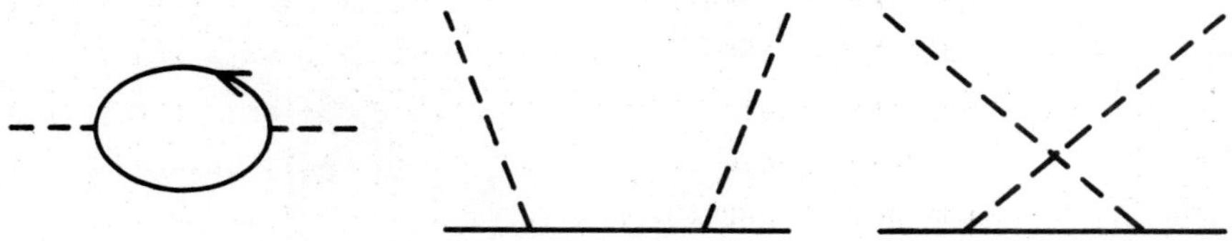

Fig. 1.1. The pion self-energy diagram, and its appearance when it is cut at one or another point on the nucleon line.

Now to put this in terms of a formula, what you want to calculate is the self-energy which I call Π. (By the way, the self-energy is defined by Dover and Lemmer in such a way that k, the wavenumber in nuclear matter is positive if Π is positive, in other words positive Π means an attractive potential).

Remembering the diagram of Fig. 1.1, where I cut the nucleon line, you will see that the scattering is essentially the derivative of the pion self-energy with respect to the number of nucleons in that state which I am considering. I have here one nucleon line and have, in fact, many nucleons that I can choose from, so the Π depends on the occupation number of the states in nuclear matter N(p), p being the momentum of the nucleon. (By the way, I find it useful to avoid confusion to always use p for nucleon and always k for pions, then you can recognise what you see in every formula). So the derivative of the self-energy with respect to the occupation number is just the forward scattering amplitude which we had before.

Now we have to look at this occupation number. What happens when I build up the density in nuclear matter? I start out by putting nucleons in the lowest momentum state, and then higher momentum states as the density goes up, so when I change the density from ρ to $\rho + d\rho$, what I add is simply occupation in those nucleon states which are just at the surface of the Fermi sea, and therefore have a nucleon momentum equal to the Fermi momentum, p_F. And so $d\Pi/d\rho$ will be an average over the direction of the nucleon relative to the pion of the forward scattering amplitude, and this average I call f_{av}. What you have then because of this derivative is something very similar to the original formula, (Eq. 1.1) from the optical theorem, only in that expression I had, as you will remember, ρ times the forward scattering amplitude, while now I have an integral of the forward scattering amplitude over $d\rho$ - not much difference.

What I have talked about so far is the scattering of a pion by a single nucleon. Of course, it is perfectly possible to have also scattering in which more than one nucleon is involved. I can have the pion come in, put one nucleon up into an excited state and then have an interaction of this with another nucleon. I make another particle-hole pair, and then the pion goes back out again. This is illustrated by Fig. 1.2.

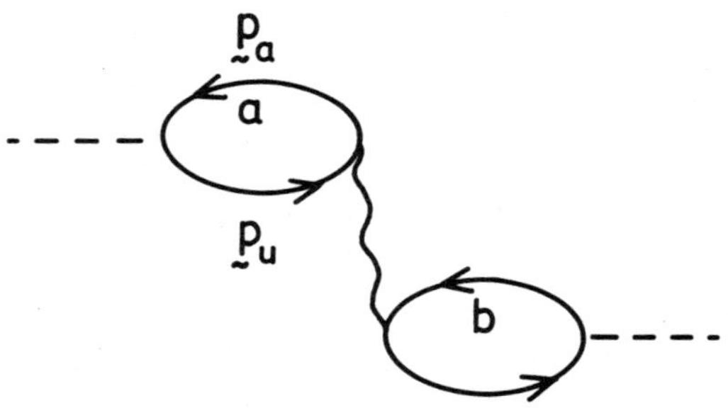

Fig. 1.2. A two-nucleon contribution to the pion self-energy; a and b are the two nucleons.

This two-nucleon scattering can, of course, be fairly complicated, but it has been shown, especially by Gerry Brown and a few of his collaborators, that the essential point in this diagram is the fact that two nucleons never can come very close to each other. There is a correlation function between nucleons which is zero at zero distance because of the strong repulsion at short distance, and once this is the case, once you have a strong anti-correlation of this type, this two-nucleon process reduces simply to the Lorentz-Lorentz factor, that is, instead of having $4\pi\rho a_1$ (this holds only for the P-wave scattering) you should divide this by $1 + \frac{4\pi}{3}\rho a_1$. So you have reduced it to a problem already solved in electrodynamics. The rest of the two-nucleon correlation apparently makes very little difference.

With this amendment then, we have essentially the same as I had in Eq. 1.4. The critical density now goes up, as you can easily convince yourself, to three-halves of what it was before and therefore just about to the density of nuclear matter. So you are still in trouble. The relation between wavenumber inside and wavenumber outside is

$$k^2 = \left[\omega^2 - \mu^2 + 4\pi\rho a_o(\omega)\right] \frac{1 + (4\pi/3)\rho a_1(\omega)}{1 - (8\pi/3)\rho a_1(\omega)} \tag{1.5}$$

I repeat that Π of course is complex because the scattering amplitude is complex, the state k of the pion always decaying by quasi-elastic scattering.

The next step is to calculate the forward scattering amplitude. Fortunately, this can be done in essentially the same way as it is done for free pions by the Chew-Low theory. In the Chew-Low theory, as you may remember, the pions are treated relativistically, but the nucleons are treated statically or non-relativistically. In the Chew-Low theory the main thing is that you have an incident nucleon, and an incident pion, and you first emit a second pion, the nucleon thereby goes into state A, then you absorb the incident pion in state B, etc., etc., until the final pion comes out. The point is now, that in the Chew-Low theory, nearly all the interaction comes from intermediate pions of very high energy, because the dispersion integral which I will show you in a minute is linearly divergent with the momentum of the pion. Therefore this pion momentum is likely to be very high and therefore the corresponding nucleon momenta are also high until your final pion is emitted. This has two consequences:

1. The nucleon intermediate states have high momenta and therefore presumably are not affected by the Pauli principle.
2. Because of the formula

$$k^2 = k_o^2 + \Pi \tag{1.6}$$

you can show that the self-energy of the pion remains finite or even goes down at high energy. Therefore as you go to high energy the momentum of the pion inside the nucleus is about the same as it is outside, and therefore the pion propagators which you have in the Chew-Low expansion will be the same as if the pion were free. Therefore, not only can you use the formalism of Chew and Low, but even the number should remain the same because the main part of the dispersion integral - namely the high energy part - remains unchanged. That's the main point of the theory.

In the Chew-Low theory you separate the scattered amplitude

$$f_{\underset{\sim}{k}\underset{\sim}{k}'}(\omega) = -\frac{2\pi}{(\omega_{\underset{\sim}{k}}\omega_{\underset{\sim}{k}'})^{1/2}} \sum_{\alpha} h_{\alpha}(\omega) P_{\alpha}(\underset{\sim}{k},\underset{\sim}{k}') \tag{1.7}$$

into an angular factor and a factor h depending on the energy of the pion. The angular factor corresponds to the various isospin and spin states. The interesting thing is the energy dependent factor, for which you get a dispersion integral

$$\begin{aligned} h_{\alpha}(\omega) &= \frac{\lambda_{\alpha}}{\omega} + \frac{1}{\pi}\int d\omega_q \; q^3 v^2(q) \\ &\times \left\{ \frac{|h_{\alpha}(\omega_q)|^2}{\omega_q - \omega - i\varepsilon} + \Sigma A_{\alpha\beta} \frac{|h_{\beta}(\omega_q)|^2}{\omega_q + \omega} \right\} \end{aligned} \tag{1.8}$$

The first term in the dispersion integral denotes the fact that you can emit a pion of a different momentum and then re-absorb it, and in doing so, you get each time a factor h_{α}, and so you get $|h_{\alpha}|^2$, with the requisite energy denominator.

Now to solve this equation, Chew and Low have shown that it is most convenient to use not h_{α} but to define a new quantity

$$g = \lambda / z h_{\alpha}(z) \tag{1.9}$$

where z is simply a generalisation, a complex generalisation, of the energy. It's just the energy taken in the complex plane. Now why is that convenient? It is convenient because the term $|h_{\alpha}|^2$ is determined from unitarity, and unitarity is expressed in the fact that h, the scattering amplitude, is essentially $\sin\delta\, e^{i\delta}$. As you know if you take the reciprocal of this, you get $e^{-i\delta}/\sin\delta$, which has the very agreeable property that its imaginary part is simply minus i and does not depend in any way on the process you are actually considering, and it is this which then permits the solution of the problem; and so the g_{α} goes to $\cot\delta_{\alpha} - i$.

I said already that you have to consider unitarity but in order to consider unitarity, it is essential that you use real values of the momentum k. As soon as you use complex values you get into all sorts of troubles and I know that very well because for about five months I struggled with just this problem: how can I put in unitarity when k is a complex number? And a complex number it ought to be from Eq. 1.6 because the self-energy Π is complex, and k_o is real, because it is $\omega^2 - \mu^2$. Well, the help for that was provided by a suggestion in a seminar at the University of Washington just about a year ago at which Professor Peierls was present. The suggestion was to make full use of the flexibility of complex variables which were introduced into the problem by Chew and Low, namely: what do you have in the complex energy plane, in the complex omega plane? Well, you have an origin, and you have a branch point at μ, the mass of the pion; then usually you take a branch cut along the real omega axis, but nobody tells you where to put the branch cut as long as it goes from the branch point to infinity. So why not make it go differently, why not make it go along the line on which k is real and omega therefore complex?

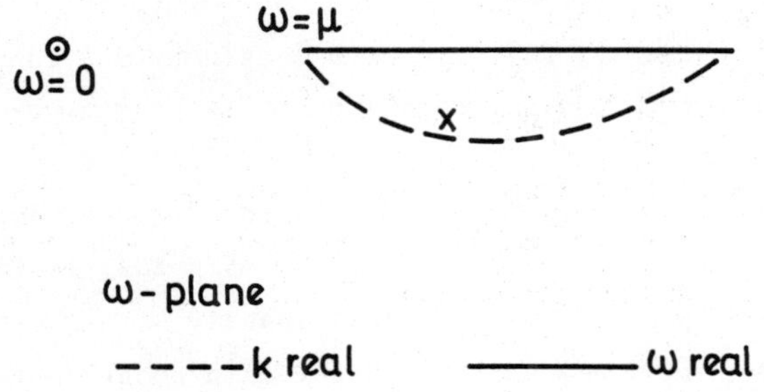

Fig. 1.3 The complex ω plane. ⊙ = origin, ------- line of real k, x = a possible value of z; solid line is branch cut along line of real ω.

I say that for real k, omega will be complex. Why do I need real values of k? This is most easily seen from the old Rayleigh scattering formula; you know that you have a plane wave e^{ikz} plus spherical waves, as you generally expand in incoming and outgoing spherical waves which have the form exp(±ikr). Clearly you get into trouble if you make k complex, either with the ingoing or with the outgoing spherical waves. So you have to choose k real to be able to define the phase shift. So I go along the solid line in Fig. 1.3 taking k real, and then I can use the fact that I know, from the phase shift expression for h, the imaginary part of the quantity zg(z), and I know this just on the positive side of the branch cut. That represents the physically permitted values which lead to outgoing spherical waves. These still have a negative imaginary part of omega but they are on the positive side of the branch cut. So you have this imaginary part because of the form of h, which is directly given to you, and is some simple expression in terms of k. Unfortunately, since the branch cut now has complex omega, you no longer get a very simple expression for what you want to know, namely zg(ẑ) is given by the expression

$$zg(z) = z - \frac{z^2}{\pi}\int_{\mu}^{\infty} dx' \left[\frac{F_\alpha(x')}{x' - z} + \frac{G_\alpha(x')}{x' + z} \right] \tag{1.10}$$

What you really would like to know is the quantity F_α in this expression and that is what normally is determined from the imaginary part of g. But unfortunately, z itself is now complex so you get really three contributions to Im g, of which only one is the thing you want to know. But then by discussing this in some detail, you can show that the other two parts are unimportant at least for high pion energy. So we then are reduced essentially to the result of Chew and Low and that, of course, is a happy solution. You have to calculate the dispersion integral, Eq. 1.10, the imaginary part of which is just what I got from unitarity, while the real part is the one which gives you the position of the resonance.

Now looking at this expression, the F_α, from all the complicated argument which I have told you, behaves about as k (the momentum) at large momenta, so that the integral in Eq. 1.10 will diverge linearly with momentum. Thus you see that indeed the high momenta give you the main contribution. Because this is so, the main contribution comes from those values of x' where I know the F_α very well - it's the Chew-Low value. Then the integral itself will be close to what Chew and Low give. The result then is g, for which I get just the same thing as Chew and Low, namely

$$g_\alpha(\omega) = 1 - \omega r_1 - \frac{i}{\omega}\left[\lambda_\alpha k_\omega{}^3 v^2(k)\mu^{-2} + \omega^2 r_2\right] \tag{1.11}$$

where r_1 is what Chew and Low call the effective range. Now the imaginary part in Eq. 1.11 is slightly modified for the reasons I mentioned, but the real part is much the same as Chew and Low obtained and therefore, it should go to zero just at the same point where Chew and Low go to zero, namely at the resonance energy. Therefore, r_1 should be just about one over the resonance energy. The imaginary part has a correction which is related to Π and is not very important. I started from going along the direction of real k, but now, by using the magic of complex variables, I can transform all this to an integral along real values of the energy, real values of omega, and then I am essentially back to where I started from with a slight correction.

So if you now want to calculate the behaviour of pions in nuclear matter, you assume the self-energy Π as known. From that you calculate g, g gives you h, and h gives you f, the scattered amplitude, and this in turn gives you the Π, that is the self-energy, and then you iterate until you get self-consistency. The final formula for the (3,3) scattered amplitude f_3 is

$$f_3 = C_k \frac{2\lambda_3}{\mu^2} \frac{k_\omega^{\;2}\, v^2(k_\omega)}{\omega(1 - \omega r_1) - i[\;\;]} \tag{1.12}$$

where k_ω is the value of the momentum which corresponds to a given energy omega, $[\;\;]$ is the square bracket of Eq. 1.11, C_k is a number, and v(k) I haven't told you. It is a cut-off factor which was introduced by Chew and Low in order to cut off the divergence of the integrals.

Now, this is all well and good, except it leaves out two important points - one is the Pauli principle and the other is some restrictions on the energy. I told you before that the Pauli principle shouldn't really come in, but that applied only to the case of the intermediate states of the pion, and it does not apply to the final state. What we have in general is that the scattered amplitude describes a scattering in which the pion initially going in direction $\hat{\underset{\sim}{k}}$ encounters a nucleon and afterwards is scattered in another direction, let's say $\hat{\underset{\sim}{k}}_f$. When it does so, then of ccurse it gives a recoil to the nucleon, and this recoil may bring the nucleon into a state which is already occupied, and this would then be forbidden by the Pauli principle. Now I no longer have the excuse that the k of the pion is high, because the final k must be the same as the initial k or less, because the pion has lost some energy. So, therefore, I do have to take into account the Pauli principle for the states of the same energy which can be reached in such a scattering by a single nucleon. (Let me for the moment consider the nucleon as infinitely heavy).

Such problems are generally done by the Lippman-Schwinger equation

$$T = K + i\int d^3k_f\ \delta(\omega_f - \omega)\ K|k_f\rangle\langle k_f|T\ Q(k_f) \tag{1.13}$$

in which you start off by knowing the K-matrix which is the matrix not taking into account the condition that you always need to have outgoing waves; i.e. K is obtained using the <u>real</u> part of the dispersion relation. You want to get the T-matrix which is the actual scattering matrix. The relation of these is given by an integral over all directions of the pions, directions of $\hat{\underset{\sim}{k}}_f$, where you go with the T-matrix to the intermediate state and the intermediate state must be empty, that's expressed by the operator $Q(\underset{\sim}{k}_f)$, the Pauli operator, and then you go with the K matrix to the final state. The condition for Q is that the nucleon momentum after this first scattering must be outside the Fermi sea so the incident nucleon momentum plus the incident pion momentum minus the intermediate state pion momentum must be greater than the Fermi momentum. Now it is useful to introduce the sum of the incident momenta, that is the incident total momentum, which I call $\underset{\sim}{P}$, and to choose $\underset{\sim}{P}$ as an axis; then you can introduce polar coordinates Θ, ϕ for $\underset{\sim}{k}_f$ and the Pauli principle simply comes down to a condition

$$\cos\Theta < \lambda \tag{1.14}$$

where λ is calculable from the given quantities

$$\lambda = \frac{P^2 + k^2 - p_F{}^2}{2Pk} \tag{1.15}$$

Well, this Lippmann-Schwinger equation, Eq. 1.13, which looks rather formidable, can be solved very easily and we have done so. The result is given in Eq. 1.16 in terms of the quantity λ, which I just defined. The solution for the forward scattering is just

$$T(\underset{\sim}{k},\underset{\sim}{k}) = -\pi K \left\{ \frac{3(1-\lambda^2)}{1 - \frac{1}{4}ik\omega K_o(1+\lambda)(2+\lambda-\lambda^2)} + \frac{1+3\lambda^2}{1 - \frac{1}{4}ik\omega K_o(1+\lambda)(2-\lambda+\lambda^2)} \right\} \tag{1.16}$$

The forward scattering amplitude contains the term

$$1 + 3\cos^2\alpha \equiv 1 + 3\lambda^2 \tag{1.17}$$

characteristic of $P_{3/2}$ scattering, as many of you will know, where α means the angle between the incident pion momentum and the incident momentum $\underset{\sim}{P}$. (Well, the main thing I wanted to show you is that it can be done, the details are not essential). What the Pauli principle will do is reduce the imaginary term in the denominator; this imaginary term I will now call damping for the sake of having a short name for it, and so the damping is not quite as big as it would be for free pions, or for an ordinary delta resonance if it were left alone.

Now the second thing that has to be taken into account is that the nucleon mass is not infinite, but is in fact very finite and is still further reduced: there is an effective mass inside the nucleus which is maybe seven tenths of tha total nuclear mass as we know from nuclear matter theory, and so that mass isn't all that much bigger than the pion mass. Now because of this the energy which may be transferred to the nucleon in the scattering may be very large, especially because there is a tendency for the pion momentum to be very large: I mentioned in the beginning that the pion momentum inside the nucleus is much larger than outside; the simple way to say it is that the pion is attracted by nuclear matter and therefore has a higher momentum inside. So since the pion momentum is large and since the nucleon mass is not very large, the energy transferred to the nucleon is quite substantial and has to be taken into account; this is another correction which works in the same direction as the Pauli principle. I will not go into this in detail. Instead of that, I want to show you the curves which have been obtained by my collaborator, Mikkel Johnson of Los Alamos, which give, according to this theory, the momentum of the pion inside nuclear matter as a function of the momentum in free space.

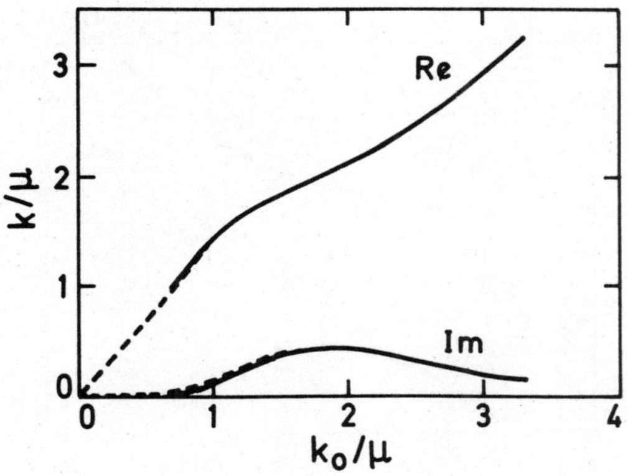

Fig. 1.4. The real and imaginary parts of k/μ, assuming density $\rho = 0.5\ \rho_o$, where ρ_o is the density of normal nuclear matter.

If I had no interaction with the nucleus I would get a straight line, $k = k_o$ for Re k, and Im k = 0. You see the momentum inside nuclear matter is substantially higher than k_o, as I have often said. If the density is equal to that of normal nuclear matter, the momentum inside nuclear matter is almost constant independent of the energy of the pion which is entering; this is an effect of all the various terms I have discussed. The inside momentum is in fact just about equal to the momentum of a pion at the resonance, and you can understand that very easily because if you are right at the resonance, then there is no attractive force and the pion should have just about the same momentum inside as outside. Let me make two more points about this. One, there is never any catastrophe. I never have any trouble from the fact that I have a zero denominator; the reason for this is the unitarity which keeps my $a_1(\omega)$ large and keeps its imaginary part large and therefore never permits the denominator in Eq. 1.4 to go to zero. The second point is that the imaginary part of k is never very large - its highest value is about $0.5.f^{-1}$, so that the mean free path of a pion inside nuclear matter is about 1f, about the distance between two nucleons. This I considered very important from the beginning. You cannot tolerate extremely large cross sections in nuclear matter, because if you did, it would tell you that the pion would get absorbed before it has seen even one nucleon, and this would be rather absurd. The cross section, in fact, is reduced to about half the resonance cross section - the effective cross section - for absorption. Fig. 1.5 gives the curve for the density of nuclear matter.

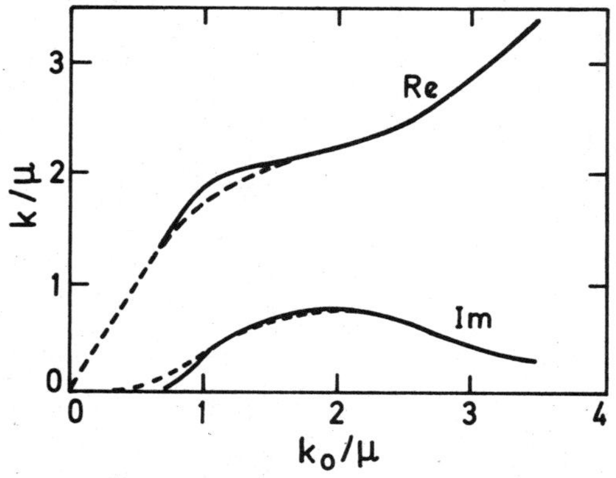

Fig. 1.5. The real and imaginary parts of k/μ, assuming density $\rho = \rho_o$, where ρ_o is the density of normal nuclear matter.

If you have lower density (Fig. 1.4), the momentum is, of course, much closer to the outside momentum, and if you have excessive density which doesn't occur in nature, then you get an interesting phenomenon, namely, because of the attraction the momentum goes up to a very high value and then drops down as it approaches the resonance.

Now what are the consequences of all this for the behaviour of pions in finite nuclei? - what I have done so far is just the case of infinite nuclear matter. First of all, while the absorption of pions in nuclear matter is not as high as an elementary theory would lead you to believe, it still is very, very strong. Therefore, when you have pions of any energy, from about 50 MeV to about 300 MeV, they are so strongly absorbed in the nucleus that the centre of the nucleus is totally black. This has two consequences - one is that the elastic scattering is done entirely by the surface of the nucleus where the density is <u>not</u> very high; and second, that from the elastic scattering you cannot learn anything about the behaviour of pions in really dense nuclear matter. The first point is very welcome because there have been many theories of the interaction of pions with nuclear matter which suffered from the syndrome which I mentioned at the very beginning and which, nevertheless, gave good answers for the elastic scattering. The reason is that <u>any</u> theory is nearly correct at low nucleon densities, and since only low nucleon density matters for elastic scattering, the older theories gave perfectly good results.

I have used the Eikonal approximation to treat this problem, and from that I find the important nucleon density which is responsible for the scattering. In the case of carbon it is .28 of the central density, in the case of lead it is .18, so roughly a quarter of normal nuclear density. That's of course very nice in many ways because it permits you to explore the surface of the nucleus, and this is one reasonable application of the pion factories, namely, to tell you more about the surface.

One use that a pion facility may be put to is to study the nucleus Ca^{48} and related nuclei. As you know, Ca^{48} has a core which is just the same as Ca^{40}, the shells 1s1p1d and 2s are filled, and then it has an outside shell of eight neutrons in the 1f shell. You want to find out where these neutrons are because electron scattering doesn't tell you. Pion scattering tells you very nicely. You go to the resonance and at the resonance the effective density of nucleons responsible for scattering for π^+'s is half the neutron density plus three-halves the proton density, for the negative pions it's the other way around, and so what I have done is to plot in Fig. 1.6 Negele's results for these densities. Curve C shows you the density of the core nucleons as a function of distance from the centre, curve f the density of the f-shell neutrons, and then curves π^+, π^- give the effective density which scatters π^+'s and π^-'s. So the nucleus should look quite a lot bigger for π^-'s than for π^+'s, about 8% bigger, and this should be quite easy to observe.

There are two very important points about the resonance which people have spent a lot of time on, both measuring and explaining - one is the resonance width; there are some papers in fact, especially recent papers, which claim that the width of the resonance is less than it is for free nucleons. This is totally wrong, the resonance width is greater for the same reason that I have explained before, namely, because k is bigger. This also makes the imaginary part of the momentum bigger, and so on. In fact, the effective

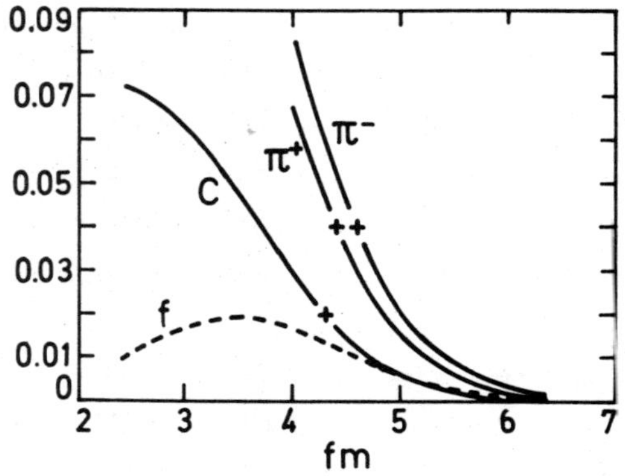

Fig. 1.6. Density distribution of nucleons in $^{48}Ca_{70}$ according to the calculations of Negele, in nucleons per fm^3. Curve C: Core nucleons (either neutrons or protons). Curve f: Neutrons in the 1f shell. Other curves: Effective density for scattering of π^+ or π^-. The + signs mark the points where the effective density is 0.04 fm^{-3}.

width can be defined conveniently by talking about the imaginary part of the momentum and setting it equal to half the imaginary part at resonance. Then the effective width is about 65 MeV, whereas for a free pion it is about 43 MeV.

Still more has been written about the position of the resonance and people have argued about how to measure the position. It seems to me clear that one should not define the resonance by the maximum of the cross section, or even by the maximum of the reaction cross section, because these cross sections are terribly influenced by such things as the finite wavelength of the pion, compared with the nuclear radius. The elastic cross section is even worse. Instead of this I think the only good definition (which has been used once or twice before) is to choose the point where the real part of k is equal to k_o, that means the nuclear scattering amplitude has no real part. This you can measure most conveniently by taking the Coulomb interference. Fortunately, Binon and collaborators, and also others, have done that, and have found that the observed energy is about 175 MeV, for this point. The free nucleon resonance is at about 195 MeV, both of these numbers in the laboratory. I have calculated from our theory that there should be a downward shift of the resonance by about 10 MeV. Most of this comes about from the effect of the small phase shifts, the (11), (13), (31) states and the s-states which you could, of course, also take into account in the free nucleon scattering, so you could compare this point to the point where the free nucleon scattering has no real part and that would be shifted down also by about 10 MeV. I cannot explain the extra 10 MeV down shift. However, I have an advantage over previous attempts, namely, all the shift numbers that I'm dealing with are of the order of 10 MeV, while most of the previous theories have numbers like + 50 - 90, etc., and it is then very hard to land at minus 20 in the end.

One interesting feature is the fact that you have a very strong refractive index for pions, in nuclear matter: the refractive index is just k/k_o. From Fig. 1.7 we see that the refractive index is about 2 at about 50 MeV, at about 130 MeV the refractive index goes down to $\sqrt{2}$, and finally it is 1 just at the resonance.

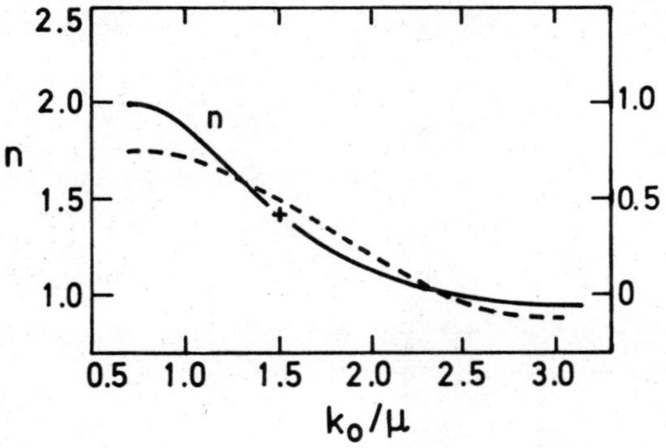

Fig. 1.7. Real part of the refractive index, at density ρ_o, as a function of k_o/μ. The + sign marks $n = \sqrt{2}$, where the probability of total reflection is one-half. ------ gives the $\cos^2$ of the critical angle for total reflection of a pion coming from the inside of the nucleus; refer to the right hand scale.

Now the refractive index means that if a pion is scattered in the nucleus to an energy, let's say of 100 MeV or so, it will have a very hard time to come out again because when it comes to the surface it will suffer total reflection. Therefore, I conclude that there should be very few inelastically scattered pions in an energy range, let's say from about 30 to about 180 MeV, and there should again be inelastically scattered pions at very low energy, because there the refractive index is reversed, because then the s-scattering, which is repulsive, becomes more important than the p-scattering.

This is a brief survey of what we know at this point about the scattering of pions by nuclear matter and by finite nuclei.

References

(1) S. Barshay, V. Rostokin and G. Vagradov, Phys. Lett. 43B (1973) 271.

(2) S. Barshay, G. E. Brown and M. Rho, Phys. Rev. Lett. 32 (1974) 787.

(3) C. B. Dover and R. H. Lemmer, Phys. Rev. 7C (1973) 2312.

Discussion

Peierls: One comment that occurs to me is that perhaps the most exciting, though not the theoretically most interesting, consequence may be that here is really a way in which one can use elementary particles other than nucleons for studying the nucleus, and where one can use high energy bombardment to get some information about the nucleus. That hope has been around since the very early days, because it is the direct extension of Rutherford's old idea of using fast particles, namely α particles, to study the atom, and he made very good use of it. Now the thought was always around that, similarly, to study the nucleus one should use particles which are much faster and which contain much more energy than the constituents of the nucleus. It never worked. There were I don't know how many beautiful schemes which looked convincing at first sight, and when you looked into all the corrections you found that the information had evaporated. Now this is something that looks different, because on the face of it you don't see that this is the way to get information about the nuclear surface - one has to go through rather sophisticated arguments which you have demonstrated, but it looks as though, when you've done all that, here is a simple tool using pions, with which to learn something about the nucleus.

Dalitz: One of these processes of absorption of pions by nucleons is a process involving a pair of nucleons. Already in the calculation the absorption is very strong, so perhaps this will not contribute very much?

Bethe: The absorption I have here of course is not absorption but scattering. But there is an additional absorption by a pair of nucleons which is very important at low pion energy and it seems to give a mean free path inside nuclei of about 20f. That's a fairly long path. However, there is the total reflection, which may occur several times, so I expect that even when I come in with a pion of 200 MeV it is quite likely that it will get absorbed in the nucleus by first being scattered and then being unable to get back out again.

Karl: One might think at first that you could scatter even from virtual pions, which you do not put in. Do you think that it would be inconsistent to include the scattering of the pion from virtual pions in the nucleus?

Bethe: Well, I don't think it's inconsistent, but it's not in my picture. I have taken strictly the point of view that any such things as virtual pions are quite marginal, and that I just look at the interaction with nucleons. Now I guess that insofar as the virtual pions are made by single nucleons each time, this process is already contained in the theory, because it is contained in the scattering of a pion by a single nucleon. So all you can get is the modification of the virtual pion density by virtue of the simultaneous existence of many nucleons, and I have left out all these effects (except for the Lorentz-Lorentz effect which I mentioned). They will have to be left to the next person who tries to struggle with this problem!

Castillejo: Can I ask what the effect of scattering of the Δ resonance on the other nucleon is? This resonance lasts for a relatively long time and it must presumably have an interaction with the nucleons. Would this affect your discussion at all?

Bethe: Well, the Δ particle you might say could have a different interaction with the other nucleons than the original nucleon. This is entirely possible - it may well be the thing responsible for the extra 10 MeV shift in the resonance energy. It is quite a difficult problem which some people have attempted to deal with, but I don't think anybody has got a very good answer for it. I have simply said that each nucleon is by itself, and you have interactions between the nucleons only in the same degree as you have in ordinary nuclear matter. So I think implicitly I have assumed that the interaction when in the Δ state is the same as for the nucleon.

2

Early Steps Towards The Chain Reaction

O.R. FRISCH O.B.E. F.R.S.

I have been invited to talk a little about that period during which Professor Peierls took part in the development of the nuclear chain reaction, in which I also participated. I cannot claim to be a pupil of Peierls, as so many here are, but I can claim to be a pupil of Genia Peierls, who taught me to wash dishes and to shave every day - a habit which, after 33 years, I have now abandoned. I shall tell you what little I remember of those days, well over 30 years ago now, when the possibility of a nuclear chain reaction was first discussed, and the first groping attempts were made to observe it or to create the conditions for it.

I had been in Copenhagen for five years, and towards the end of that period, in 1938, the general situation in Europe, with Hitler's activities, appeared to be getting more and more unstable; and even though I don't usually show much political foresight or initiative, on that one occasion I did. When Professor Oliphant came over on a visit from Birmingham, I told him that I felt worried about Denmark being occupied by the Germans, even though one of my friends had said, "Why should Hitler occupy Denmark? He has a telephone, hasn't he?" Anyway, I felt that it might be unsafe, and that if it came to any sort of war effort, I would prefer to be on the other side. He then very kindly invited me to come over in the summer of 1939 to discuss matters with him, and this I did in June with two small suitcases, leaving all my other belongings behind. Before we had been discussing very long, war was declared and I had to live out of those suitcases, and on what I gradually acquired.

At first I stayed in a boarding house in Birmingham but later, I think it was early in 1940, the Peierls invited me to stay at their house and that is where my education was successfully taken in hand.

Already when I came to Birmingham I was aware that fission was not just one more nuclear reaction, liberating an unusually large amount of energy, but something much more exciting, and certainly the possibility of a chain reaction had been widely discussed. The first person who mentioned it to me was Christian Møller in Copenhagen, who had worked a bit with Bohr on the theory of excited atomic nuclei and what sort of energies were needed to liberate a neutron. He came and said that he thought that in the creation of two fission fragments, enough energy must be left inside each one, apart from the kinetic energy of the whole fragment, to give a neutron a good chance to escape, so that probably two neutrons, or perhaps even more, might be liberated, and that would make a chain reaction possible. And I remember that I felt very bright when I immediately said, "Ah, now that's impossible because then uranium deposits would not exist! They would have blown up long ago!" But it took me less than a day to realise that this was a feeble argument: uranium deposits are small in extent so that neutrons can escape, and they contain all sorts of impurities which might absorb neutrons; later, of course, it became clear that uranium 238 was itself an "impurity" of that kind which didn't help with the chain reaction. In fact, during the time when I was still in Copenhagen, Niels Bohr had suggested that the observed fission caused by slow neutrons in uranium was not due to the bulk of it, to the U238, but to the light isotope U235, an idea which seemed very odd to most of us because there was so little of it - less than 1% - and the activity of fission products created by slow neutrons was quite comparable with the additional activity created by fast neutrons. So the fission cross section had to be quite large. On the other hand, large cross sections were known, although, of course, at that time the results which had shown cadmium and other substances to possess large cross sections were still fairly new, and I have a feeling they hadn't sunk in - we were always still surprised and regarded cadmium rather as a bizarre sort of nucleus. Anyway, it was an important question; some quite distinguished physicists didn't believe Bohr - I remember that Fermi didn't think that Bohr's arguments were strong and he didn't think that uranium 235 was the only isotope undergoing fission with slow neutrons. And so when I came to Birmingham, although my main job really was to help answer students' questions which came up after Oliphant's lectures, I did want to do some research and I thought it might be worthwhile trying to separate quite a small amount, or even just enrich quite a small amount, of uranium sufficiently in U235 to tell whether the effect of slow neutrons was thereby increased.

Looking around the literature, I found that the method of thermo-diffusion, invented by Clusius, was comparatively simple to set up, and I remember that I struggled for weeks with the mathematical theory of a Clusius tube in order to figure out what was the best geometry. I persuaded Oliphant to let me have the use of a small lecture room to set up my experiment. Now the reason I had to set it up in a lecture room was that by that time the whole Institute was taken over by some sort of war work which I didn't know about - that is to say, I really didn't know about it, I didn't just pretend not to know about it. It was, of course, radar. All foreign-born people were at first excluded from that work, and I remember one scene in the tea room when Oliphant, tea cup in hand, wandered up to Peierls and casually said, "Peierls, if you were asked what are the resonances of electromagnetic waves in a hemispherical cavity, do you think that could be calculated?" and Peierls solemnly replied, "Yes, I suppose one could do that", and went away and did the calculation and never let on that he did know what it was all about. Well, I think after a while I also tumbled to what was going on. Anyhow, the facilities were limited, and for my experiment I was given a small lecture room and a student. The student was Ernie Titterton (now Sir Ernest) who later followed Oliphant to Australia and became head of the Physics Department at the University of Canberra.

I wanted to mention Ernie Titterton because many people don't know that he made quite an important discovery, which was spontaneous fission. That came about because I felt at one stage that we ought to make experiments to measure fission cross sections for slowish neutrons with the sort of intermediate energy which Bohr said would not affect U238. To separate uranium isotopes might take a long time, and I hoped that in this way we might be able to study the effect on U235 of the kind of neutrons which we might expect to be operative in an atomic bomb. For this we needed an ionisation chamber containing a fair amount of uranium in a layer thin enough to let the fission fragments escape. That was an ionisation chamber constructed like a condenser with a number of alternate plates separated by a few millimetres of air. It naturally had a fairly large capacity, but that didn't matter because the fission pulses were so large that, even with a large capacity, they were not going to be drowned by noise. That chamber was duly constructed, and Titterton built most of the electronics to run it, and I kept complaining that his electronics had loose contacts or something, because we kept getting occasional pulses for no visible reason. He insisted that his electronics were sound - we did all sorts of tests to check it - and I remember we spent several days minutely searching every square inch of the room to make sure that some neutron source hadn't accidentally been left there. In this way it took quite a while before I became convinced that Titterton was right and that uranium suffered fission spontaneously from time to time. The rate was so low - I think it was something like a few pulses per hour - that it was quite difficult to have reliable statistics in a short time. But of course Titterton couldn't publish this discovery in wartime.

The other problem consisted in obtaining a suitable neutron source, and for this I decided to use a source of photo-neutrons and the best way to make a sufficiently intense source was to get a strong preparation of radon and surround it by beryllium. Some enquiries led me to the Christie Hospital in Manchester, which had a strong source of radium, but that source had been removed from Manchester, for safety. It was in Derbyshire at the bottom of a cave, known as the Bluejohn cave because they mined that sort of semi-precious stone there. I don't know through whom and how this was arranged, but one day I was told to go to Manchester and from there was taken by car to the Bluejohn cave. We climbed down endless ladders and went through very muddy corridors into a room in which a crude wooden table supported a perfectly normal equipment - the usual glass forest - for milking a radium source and filling a little capillary with radon. That radon source, with a small amount of lead around it, was put into my suitcase, and I travelled back to Birmingham with the suitcase either in my hand or in my lap, or else in the luggage net above me - 700 millicuries of radon. It doesn't seem to have done me any lasting harm. Nor do I seem to have suffered any harm from the fact that since radon decays fairly fast, I decided to work 36 hours non-stop in order to use the radioactivity while it was fresh, and we got enough fission pulses out of our chamber to get an estimate of the cross section.

Peierls and I had already written that famous report which led to the forming of the Maud Committee and in general to the beginning of the work on the atomic bomb in this country; it can all be read up in a white paper which appeared after the war. The way this came about was again an intriguing complex of coincidences. Perhaps one should not call it a coincidence that

Peierls had worked out a formula for the critical size of a mass of fissile material, a formula which of course contained the then unknown fission cross section. Such a formula, or rather a rough one, had already been worked out by Francis Perrin, and Peierls had extended that formula to the case when the material itself possesses absorption, and I think he had already included the surrounding reflector in his calculations. All this merely seemed to show that even if uranium was sufficiently fissile you would have to use very large quantities, measured in tons. But where the coincidence came in was that I was asked to write a summary of nuclear physics for a report annually published by the Chemical Society. And in that summary I mentioned, of course, the discovery of fission, which had happened in the year under report, and I also mentioned that although there existed the possibility of a chain reaction, it was not yet clear how such a reaction could be achieved, and, in particular, that Niels Bohr had advanced very good arguments - which I quoted - that such a reaction could not lead to an explosion of any significance. Bohr had pointed out that uranium 238 would probably not support such a reaction, because neutrons were much more likely to be slowed down by inelastic collision and absorbed than to cause fission, and uranium 235 was present in much too small quantity. The only chance he could see of creating the chain reaction would be to add some sort of moderator, some sort of slowing-down material, which would bring the neutrons down to the very low energy where they were likely to cause fission in U235, but that process would take times of the order of a sizeable part of a millisecond, and for the whole chain reaction to develop would take several milliseconds; once the material got hot enough to vaporize, it would begin to expand and the reaction would be stopped before it got much further. So the thing might blow up like a pile of gunpowder, but no worse, and that wasn't worth the trouble. All this I put into my report on fission, which then afterwards of course turned out to be a very useful piece of camouflage, as Margaret Gowing has pointed out in her book, because nobody would suspect someone who had convincingly proved that one can't make an atomic bomb, of going and making one, or trying to.

However, by writing this article, I had been made to think about the problem much more thoroughly than before. I had become aware of Peierls' calculation and I asked him for the formula, and just sort of playfully I put in an estimated cross section for U235, and as probably most people here know, the result was a critical mass of the order of a pound or so. I still don't know, and I wonder if Peierls remembers, where we got that estimated cross section from - we used 10^{-23} cm^2 - which was a very fortunate over-estimate, because if we had put in the true figure, the problem would have looked much more difficult and might never have been tackled here. So, as so often, optimism has its uses.

Let me now quickly finish what I was saying about the Clusius tube attempt. The tube which I built in Birmingham never worked. It was made of glass and heated from inside while the outside was cooled by water running down it. The water insisted on running down one side and the tube curved horribly - though, oddly enough, it never broke. Before I ever got it to work - and I would only have tried in the first place to separate the isotopes of air - namely oxygen and nitrogen, to see if that would work - the whole direction of the work was changed. But my familiarity with the theory of the Clusius tube allowed me to estimate how many Clusius tubes might be needed - if the separation constant was favourable - to separate the order of a pound of uranium 235, and it turned out it was quite a small project, considerably cheaper than building a battleship, shall we say. I think it was Oliphant who encouraged us to put this down in a report which was sent to (Sir) Henry Tizard. He in turn got other people interested and the thing snowballed into the Maud Committee and a sizeable effort, which was finally transferred, or largely transferred, to the U.S.A. where the atom bomb was eventually built.

I did finally get a Clusius tube to work in Liverpool, where I was transferred to in the middle of 1940 (I think), when it became clear that I would want bigger resources and that it was too difficult to do that work in Birmingham. That tube gave a separation coefficient of zero, which of course at first made me suspicious that I had messed up the experiment somehow, but in fact the coefficient was indeed found by others to be very much smaller than the value we had put in our estimates, so that the method of thermo-diffusion - the Clusius method for isotope separation - would not have been a practical one. Very fortunately I believe Peierls had never quite believed in it and had started early on to work out the theory of a method that was bound to work, diffusion through porous membranes, but I think Dr. Berman will tell you more about that aspect of the work.

As to the Maud Committee itself, I remember practically nothing of that intrinsically very interesting period in my life when, for a while, despite being technically still an enemy alien, I was engaged in some of the most secret work. I was transferred, as I said, to Liverpool, where I believe I was the only enemy alien because it was a prohibited area. Fortunately, almost everything of that kind was remarkably lax in those days. For instance I had a friend in North Wales, whom I visited frequently, and I usually spent several hours in Chester, a very attractive old town; only many months later did I discover that Chester was a different prohibited area, for which I had no permit at all! Sometimes when I came back from Chester, the bus would stop outside Liverpool because there was bombardment going on, and the driver would say, "We don't go any further - all get off!" and then we would have to get lifts into town. I remember riding into Liverpool in army lorries in the middle of the night when I was under curfew. So, life was tolerable in spite of restrictions which would have been very difficult if they had been strictly obeyed. It's also an odd fact that we took no security precautions whatsoever. We had no safes, we had no guards or anything like that, and yet it was a perfectly guarded secret - nobody has ever discovered a leak in our work at the time when I was in Liverpool. Rotblat gave lectures in which, among other things, he spoke about fission. He neither put any particular emphasis on it, nor did he try to keep its existence and even the possibility of a chain reaction secret. The supreme example of this type of secrecy was perhaps one little episode that happened, I believe, in the summer of '41 when Charles Lauritsen from California came over to see what we British were doing. His main interest, I believe, was in rocketry and proximity fuses, but he came and attended one of the Maud meetings. Afterwards I walked home with him to his hotel, and there on the crowded highway of Piccadilly we were talking about everything concerning the atomic bomb, possibilities of isotope separation, cross sections, whatever you like - in Danish!

Discussion

Peierls: Could I just answer the question you raised, about where the estimate of the cross section came from. I remember that quite clearly, because we had read the paper of Bohr and Wheeler and had understood it, and it seemed to convince us that in those circumstances for neutrons in U235 the cross section would be dominated by fission, and that scattering and inelastic γ-ray capture cross sections and so on would be negligible. This turned out to be incorrect, but I think that, if you took the ideas of Bohr and Wheeler literally, it followed. Then also you could reasonably assume that the total cross section for not too low energies would be the geometric cross section of the nucleus - if a neutron hit the nucleus something was bound to happen. And putting those two things together we got the number that we used. I think the number that we got was not just one pound but a few pounds, and we were not really very far out in the number. This was because of a cancellation of two errors - one was the overestimate of the cross section, which you mentioned, the other was that I had not bothered to consider the possibility of surrounding such a device by a reflector (the formula I had worked out related to a bare Uranium sphere), and those two errors very largely compensated so that our number was too small, but not wildly so.

Perhaps I could add one recollection to this point on security which you mentioned. Indeed, we thought in those days that of course it was all very important and one shouldn't spread it, but the best way of not spreading it was not to attract any attention by safes and sealed envelopes and what not. I remember we were writing our memorandum for the government together in my room in the Physics Lab on the ground floor; it was a fine day and the window was open - it opened on to a garden or an open space - and while we were discussing the wording a face suddenly appeared in the open window. And we were a little worried! It turned out that just underneath that window (which was facing south) people were growing some tomato plants, and somebody had been there bending down inspecting what these plants were doing. I don't think he heard anything of what had been going on in the room, but it was a bit worrying!

Kurti: Professor Frisch mentioned the fact that during the period, early in the war, which he spent in England, he was an enemy alien. Now that is a very interesting sequence. Of course, he couldn't have easily become a British subject, because in those days, just as today, you had to have been in the United Kingdom for five years. However, when in 1943 he was going to the United States, it was decided that it would be wrong for the British Government to export an enemy alien to the United States, to help with the war effort. So he got the naturalisation and everything, but when he arrived the American immigration officers got very suspicious, and they really thought he was an impostor, and a very stupid impostor! Because they looked at the Passport, a British Passport, and the first thing they saw was "British Subject by Naturalisation: Naturalisation Certificate issued 23rd November 1943". Then they looked at the Passport: "British Passport issued 23rd November 1943". Then they looked again: "British Exit Permit issued 23rd November 1943". And then to crown everything: "U.S. Visa issued 23rd November 1943". Fortunately, he did get out of the difficulty!

Frisch: Yes! I don't recall that incident - I'm very glad you've told me about it. I suppose I never heard about it at the time! It was all done behind closed doors, as it were. But I certainly do remember how, something like two weeks previously or so, Chadwick one day told me, "Would you like to work in the U.S.A. for a bit?" and I said, "Yes, I would enjoy that." And he said I would have to become a British citizen, and I said I would enjoy that even more. Then, about two days later, a police officer came to take down my data - date of birth, parentage and so forth; and when he left he said, "You must be a damn important chap - I've been told to get you through the mill in one week. I hope you'll remember me if ever I am in trouble." I also remember the day when I had been told to come to London with my luggage, all I wanted to take, and then a sort of ballet was performed in which people kept coming into that office in Old Queen Street, and there was one lady secretary who immediately gave us our next marching orders. I remember going out to a notary to swear the oath of allegiance, then going somewhere else to pick up a passport, then going somewhere else to be relieved from military service, otherwise I would have been a deserter, technically. Then I had to go to the American Embassy where somebody was waiting with a stamp to put it into my passport. I went to the Censor's office, who looked at my books and papers and was very interested to find that I had some Danish books, from which he started to read, because he happened to know Danish - an odd coincidence. Then I travelled through the night to Liverpool, and the next morning we embarked - there was some difficulty in getting me embarked because I had forgotten the piece of paper which entitled me to go on the ship, but fortunately the leader of the party, I think it was Wallace Akers, was present, and he told the guard, "That's all right, he belongs to the party" and shoved me through. And then we lay in the harbour for another day or so, eating oranges and fried eggs and other such unheard of delicacies, just to mislead any enemy spies who may have been lying in wait with submarines - because we didn't travel in convoy, we zig-zagged across the Atlantic; we were never told where we were until we finally landed.
Weisskopf: I would like to add another anecdote. At that time I was in Los Alamos, and I was told by the Authorities to go down to the railroad station in Santa Fe "because three British gentlemen are coming". I was slightly surprised why just I should go to meet three British gentlemen, and I was also slightly worried because my English was even worse then than now - it was enough to talk to Americans, but I was sort of worried about meeting three

British gentlemen. And how great was my astonishment when the door of the railroad car opened and out came ----- Otto Frisch, Simon and Bretscher!

3

S&U

R. BERMAN

If your eyesight is bad, you might have thought from the title that I was going to talk about symmetry in elementary particle theory. But in fact the "S" and the "U" are quite separate, and refer to the two bits of physics in which I have made contact with Professor Peierls' work.

The "S" refers to the work I did when I first came to the Clarendon Laboratory during the war. I was immediately shown out into the yard - into the building which now houses the hydrogen liquifier - where there was an apparatus for measuring flow through and across porous barriers; this was, of course, to be the basis of the gaseous diffusion method for isotope separation. It's only more recently (see Eq. 3.1) that I've realised that there's no point in having complicated apparatus to see whether Peierls' calculations are correct or not.

"S" is the symbol used for the separation efficiency of these barriers; a last piece of work that my research group in the Clarendon completed was to do with the anisotropy of Umklapp processes, and that is the "U": - to their friends Umklapp processes are known as U-processes and, of course, this is another invention of Peierls - something he did as his Ph.D. thesis.

During this 32-year period, I think the main thing I have achieved is to formulate my golden rule:

$$P = \pi|\langle 2|V|1\rangle|^2 \rho(E)t$$
$$\rightarrow 0 \text{ as } t \rightarrow \infty \qquad (3.1)$$

You'll notice that it differs from Fermi's by $2/\hbar$. The interpretation, however, is a bit different and in fact I think it's more accurate in its predictions. What it says is that the probability P that, after time t, any theory formulated by Peierls will have to be modified from its initial formulation 1 to any other later formulation 2 as a result of the perturbing effect of a high density of experimentalists $\rho(E)$, goes to zero as t gets very large. The reason is obvious - the theory is always correct straightaway, but the experimenters are not always very good at getting the right answer first time. If I were being frivolous - which I'm not - I would point out that if this is a complex expression, then we have to take the real part of it, and we put Re in front of the π.

If one reads certain textbooks on spectroscopy - and I'm glad to see that the author of one I have in mind is not here - one might think that Peierls' first work on isotopes was in 1931, because a number of books refer to a paper by Peierls and Pauli on isotope shift in spectral lines. If you actually look up the given page in Physikalische Zeitschrift you find a paper by Esterman, Frisch and Stern; as Frisch is included, it's not too inappropriate for this meeting, but you don't see the name Peierls, and it's not on isotope shifts in spectral lines, anyway. And I gather that Peierls never worked on this subject and indeed has never written a paper together with Pauli. So you can cross it out of your textbooks. However, Peierls has done some real work on isotopes, to which reference has already been made, and his contribution to the atomic energy project was really so great that it's quite impossible to refer to it even fractionally and still get to Christ Church in time for dinner. Fortunately, Professor Gowing has written the history of the British Atomic Energy Project during the war, and one can see from there the enormous contribution that Peierls made. If one wants a very crude estimate of his importance, one can look in the index and see how many references there are to him. There are far more to him than to anyone else at all - more than Churchill - more than Attlee - more than Roosevelt - and you can only beat Peierls if you add these three rather minor figures together, which perhaps gives us some idea of what he did. All this started, as Frisch has pointed out, with the Frisch-Peierls memorandum which really woke people up to the possibility of making an atomic bomb. In this memorandum there are the nuclear physics aspects, something about separation, for which, as Frisch has said, at that time the thermal diffusion method was favoured, then things about the explosive power, the hazards, the long term problems, and so on. If one goes and looks at, for example, the list in the Public Records Office which Mrs. Gowing obtained for me, the number of reports written by Peierls is quite extraordinary, and one just can't imagine how he had time to do all this without working a minimum of 24 hours a day.

We do, in fact, have evidence that he was not working all the time, which of course he should have been at a time of national emergency. We don't have any incriminating tape recordings which, naturally, weren't made in those days, but on the other hand, the typewriter had been invented. Here is the start of a report on Peierls:

Peierls, Dec. 13th

Make appointment for middle of next week.
is at home, except on Tuesday and Friday in evening

From the material in it, the year must have been 1940, when the calendar was exactly the same as this year. If you look the days up, you will see that the second evening, the Friday, when he was not working but was going to be out (enjoying himself, presumably), would have been only December 20th: really rather a long time before Christmas to be taking two evenings off. Incidentally, for those of you who were in Physics before 1956, I can show the end of the report: whether what goes in between the beginning and the end is still secret I never know. (What is supposed to be a secret from whom is rather difficult to find out). Here are the last few lines:

Ask him to calculate
these things more in detail.

ln 2 affair.

and I think people who knew one of the other giants of the British Atomic Bomb Project would recognise this sort of phrase, asking him to 'calculate these things more in detail.' That's very characteristic and the other thing, the interest in ln 2, does perhaps suggest someone whose interest is in thermodynamics, and you may recognise the handwriting as well.

From my own experience, which is very small of course, I am not aware that anything Peierls did, in those days, turned out to be incorrect. There was just a small chance I might have found something with Professor Kurti's help: in his usual wicked way he suggested I should look at two papers which differed in time by two years, both on the equilibrium time of a separation plant. And, certainly, when I looked at the second one, it did look as though there might have been a mistake. I'll just quote the beginning: "The purpose of the present report is to record the position as regards the equilibrium time of an ideal plant, i.e. a plant arranged for optimum operation, on which there are somewhat misleading statements in our previous reports." Well, I didn't go through all the mathematics, which I am sure I'd be quite incapable of understanding anyway, but it did seem to me there was nothing really wrong with the calculations: it was just that the starting conditions assumed were not perhaps as appropriate as would have been useful in practice, but from that point on the calculation, I believe, was probably all right.

As I have said, I could not possibly do justice to this aspect of Peierls' work, so I can only recommend you to read about it in Professor Gowing's book. I must leave that part of my subject with those very few remarks, and go on to the "U" part, which is the "Umklapp" process, invented in Peierls' Ph.D. thesis in 1929. I don't know what other people were doing that year, but I remember distinctly that during that summer (the paper was written in July of that year), quite unaware of the experimental work being prepared for me in Zurich, I was on the beach learning how to build sandcastles and make holes through them so that balls could somehow go on spiral paths inside as well as outside. Until I looked at the paper again recently for this talk, I thought that Peierls' main achievement in it had been the invention of a word with three consecutive p's. However, as every first-year undergraduate thinks he understands, owing to the uncertainty principle, if one looks too closely, so that Δx is small, there is a corresponding great uncertainty in the conjugate coordinate p. Fig. 3.1 shows a typical result of examining this great paper too closely. Here $\Delta x = 2.6$ cm in the original, and you can see that Δp is 1. In fact, in the paper as a whole, the word appears four times with three p's and four times with two p's, and perhaps as Professor Peierls is here, he may be able to tell us what the true spelling is. Last week when I was talking to him, Peierls tried to persuade me that the oral examination for a Ph.D., in his day, was more searching than it is in present day Oxford. This may very well be true. We don't have such brilliant research students for whom we can give very difficult oral examinations, but we do make up for it by paying attention to the spelling that goes into the thesis.

The "Umklapp" paper concerns the heat conductivity of crystals, and the der Entropie bei festem Wert von J, und dann erst der Wert von J durch die **Umklapprozesse** langsam geändert wird.

Analytisch drückt sich das so aus, daß wir die linke Seite von (68), die wir uns als Operator, der auf die Funktion λ wirkt, vorstellen wollen, in zwei Teile zerlegt denken:

(69) $$D = D^0 + D^*,$$

wobei der Operator D^0 alle die Glieder umfaßt, bei denen keine **Umklappprozesse** stattfinden: genauer gesagt, alle die Koeffizienten $D_{FF'F''}$, für die

Fig. 3.1

previous theory of some consequence had been developed fifteen years earlier by Debye. He had treated a crystal as a continuum, as in his specific heat theory, and he came up with the result that at normal temperature, the thermal conductivity of a pure crystal should be inversely proportional to the absolute temperature, and since that day any theory, however crazy, has always produced this answer for normal temperatures. The first thing Peierls showed, however, was that if a pure crystal could really be treated as a continuum, then a correct calculation would lead to a conductivity which would always be <u>infinite</u>, so that Debye's result was really not true for a continuum model.

To summarize briefly what went into this 1929 paper, I will actually use the term phonons for the carriers of heat, the quanta of lattice vibrational energy, although that term was only invented three years later in 1932 by Frenkel. Fig. 3.2 illustrates the difference between normal phonon processes and U-processes. Figs. 3.2(a) and 3.2(b) both show a section of a cubic Brillouin zone. In an (anharmonic) interaction involving three phonons with wave vectors $\mathbf{q}_1$, $\mathbf{q}_2$ and $\mathbf{q}_3$, we can have a process, for example, in which phonons $\mathbf{q}_1$ and $\mathbf{q}_2$ interact and give rise to a phonon $\mathbf{q}_3$. If this $\mathbf{q}_3$, obtained as the vector sum of $\mathbf{q}_1$ and $\mathbf{q}_2$, falls within the Brillouin zone, we have what is now called a normal process, shown in Fig. 3.2(a). By themselves, normal processes do <u>not</u> lead to thermal resistance. If one wants to be very crude - I think Professor Peierls will have to pay no attention - one would say that the same amount of energy is being transported in the same sort of direction before and after the interaction; before the interaction it's being carried

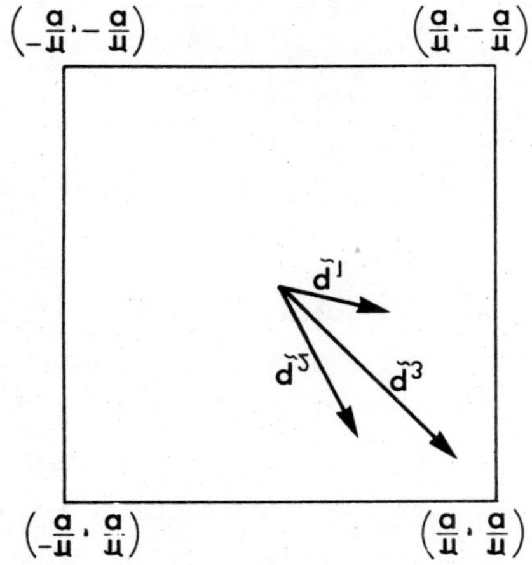

Fig. 3.2(a)

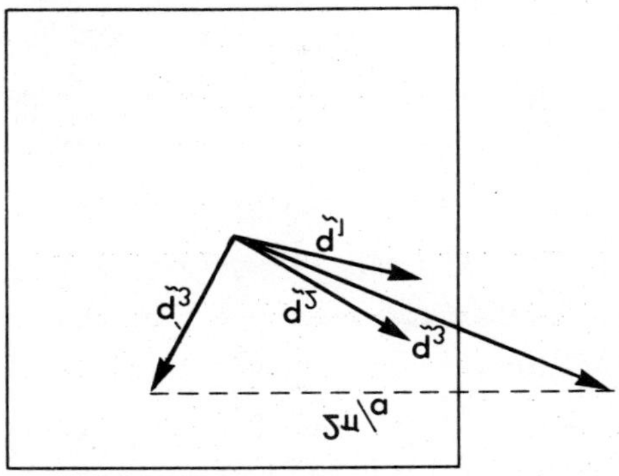

Fig. 3.2(b)

by $\underset{\sim}{q}_1 + \underset{\sim}{q}_2$ and afterwards by $\underset{\sim}{q}_3$ in more or less the same direction. If, on the other hand the vector sum of $\underset{\sim}{q}_1$ and $\underset{\sim}{q}_2$ falls outside the Brillouin zone, then to find the equivalent phonon which is the result of this interaction, we either add to or subtract from the resultant a reciprocal lattice vector to bring it back inside the zone (Fig. 3.2(b)). We then have a phonon which is travelling in the opposite direction to the initial phonons; this is what Peierls called an Umklapp process, and it does give rise to thermal resistance. The thermal resistance depends on the probability of such processes occurring and Peierls showed that at 'normal' temperatures, the thermal conductivity does indeed come out to be inversely proportional to the absolute temperature, the result which Debye had obtained, but incorrectly. This result was also more or less what the experiments at the time suggested and is also more or less what present day experiments give.

In order to get the behaviour of the conductivity at low temperatures, about which nothing was known at the time, Peierls made the following fairly simple argument. At low temperatures, a typical phonon wave vector, let's say $\underset{\sim}{q}_1$, will be fairly small. In order to get the resultant of a collision between $\underset{\sim}{q}_1$ and a second phonon $\underset{\sim}{q}_2$ to lie outside the Brillouin zone, $\underset{\sim}{q}_2$ will have to be rather large, having an energy approaching the maximum possible, $k\Theta$. So what we need to know is the probability of there being such large q phonons. At low temperatures the probability is $\exp(-E/kT)$ where E is the phonon energy. Peierls took E to be some appreciable fraction of the maximum energy possible, setting $E = k\Theta/b$ where $b \approx 2$. The thermal resistance should then be proportional to $\exp(-\Theta/bT)$. For the conductivity we just invert this, and then as we've done a rather rough calculation, we say that there will probably be some powers of the temperature in front, giving the result

$$\kappa \propto T^a \exp(\Theta/bT) \tag{3.2}$$

At low enough temperatures the exponential will dominate, so one will get an enormous departure from the T^{-1} behaviour. At the time, there was absolutely no experimental evidence for such a violent departure from the simple T^{-1} law. But sure enough, by my golden rule Eq. 3.1, when sufficient experiments had been done, it was found that Eq. 3.2 was, indeed, true.

One might think of establishing by experiment the values of a and b which were left vague by Peierls. If we plot $\log(\kappa T^n)$, where n ($\equiv -a$) is some power, versus T^{-1}, then when the right value of n (or a) is found, we should get a straight line, from which a and b (knowing Θ) can be determined. In fact, this approach doesn't get us very far. Fig. 3.3 shows a plot of log (κT^n) versus T^{-1} for sodium fluoride, and it can be seen that a straight line fit is equally good for n = 0, 1, 2 or 3. This means that from experiment we can't tell what a and b are. We can couple them together: for each curve the pair (a,b) have certain values, but we can't find one unique pair. The four straight lines can be represented by

$$\kappa \propto \exp(\Theta/3.5T),\ \kappa \propto T\exp(\Theta/2.9T),\ \kappa \propto T^2\exp(\Theta/2.4T),$$
$$\kappa \propto T^3\exp(\Theta/2.2T)$$

Furthermore, we could have chosen fractional powers of a, not just integers. The exponential is so dominant that we just cannot determine unique values of a and b. The same thing is true for lithium fluoride, where we find that we can represent the results by any of

$$T^{-1}\exp(\Theta/6.0T),\ \exp(\Theta/4.1T),\ T\exp(\Theta/3.3T),\ T^2\exp(\Theta/2.6T),$$
$$T^3\exp(\Theta/2.4T)$$

The same holds for other pure crystals.

This prediction of the rapid rise in conductivity at low temperatures was the main result of Peierls' 1929 paper and Ph.D. thesis, but it is interesting that in just a few lines at the end of the paper there are some remarks about two other things which have occupied quite a number of Ph.D. students at intervals since then. The first comment concerns point defects. One assumes

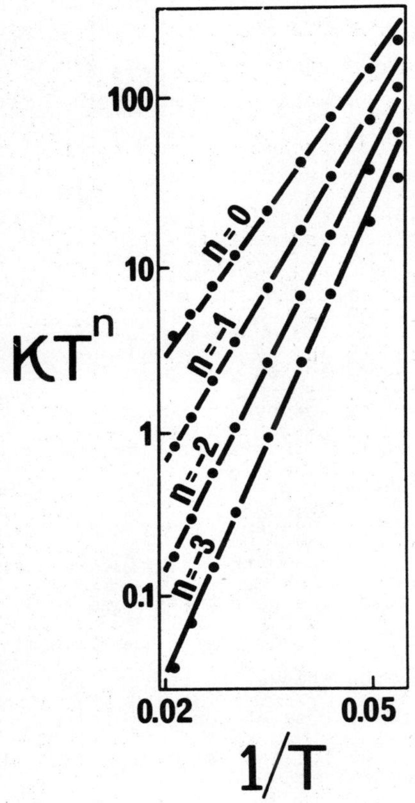

Fig. 3.3. Thermal conductivity of pure NaF multiplied by different powers of T, as a function of 1/T.

that point defects scatter phonons according to the Rayleigh scattering law ($\propto \omega^4$). Peierls indicated that if only point defects are important, as they might be at rather low temperatures, where the Umklapp processes have died out altogether, then if one puts this scattering law, and the corresponding mean free path and so on, into the expression for the conductivity, the appropriate integral over all phonons diverges. It would thus appear that at low enough temperatures, point defects should have no effect on the conductivity. Peierls pointed out - although he didn't go into as much detail as for the Umklapp process - that if anharmonicity is taken into account, then this infinite conductivity does not occur. Exactly how this divergence is removed is still pleasantly occupying Ph.D. students to this day.

The other thing which Peierls remarked on was that at relatively low temperatures in an ideal crystal, the phonon mean free path gets so large that within the crystal there is no scattering, and the phonons are only scattered at the external boundaries of the crystal. It turns out that for a rod-like specimen the heat flow is proportional to the cube of the radius, so that if we divide by the cross sectional area to get a 'conductivity', κ is proportional to the radius. Peierls pointed out that at low temperature when the mean free path is so large, it is not possible to define a true conductivity which is a unique property of the material. This, again, has given work to Ph.D. students and Fig. 3.4 shows this effect.

These measurements were made on very pure lithium fluoride from which even the lithium-6 isotope was removed. The part of the curves above $T \approx 20^{o}K$ confirms the result of Peierls' thesis that the conductivity rises very rapidly as we go down in temperature; this is the part that is represented by an exponential and it is a true conductivity - you get the same answer whatever the size of the specimen, if you do the experiments correctly. On the other part of the curves, the mean free path has become so large that it is limited by the size of the crystal. In this region the quantity one may call the conductivity is strictly proportional to the radius: just as with the Umklapp processes and the exponential rise of conductivity there was, at the time, strictly speaking no evidence for this size-effect at all. There were some misleading experiments by Eucken, and I think Peierls was not a good enough experimentalist to realise that Eucken's results were absolutely meaningless. So that was another prediction (not worked out fully, but lesser people can do that) just produced in a Ph.D. thesis.

I think my time is just about exactly up and all I have done is to summarise a few things that Professor Peierls did in his youth. I can't really summarise such a brief summary, but if I can express to Professor Peierls the spirit of what I have said, it is that on behalf of everyone in the Clarendon Laboratory, we hope you will still come and have tea and coffee with us, so that in just a few seconds over a cup of tea you will continue to explain to us what we are trying to do in our experiments, which we don't always understand.

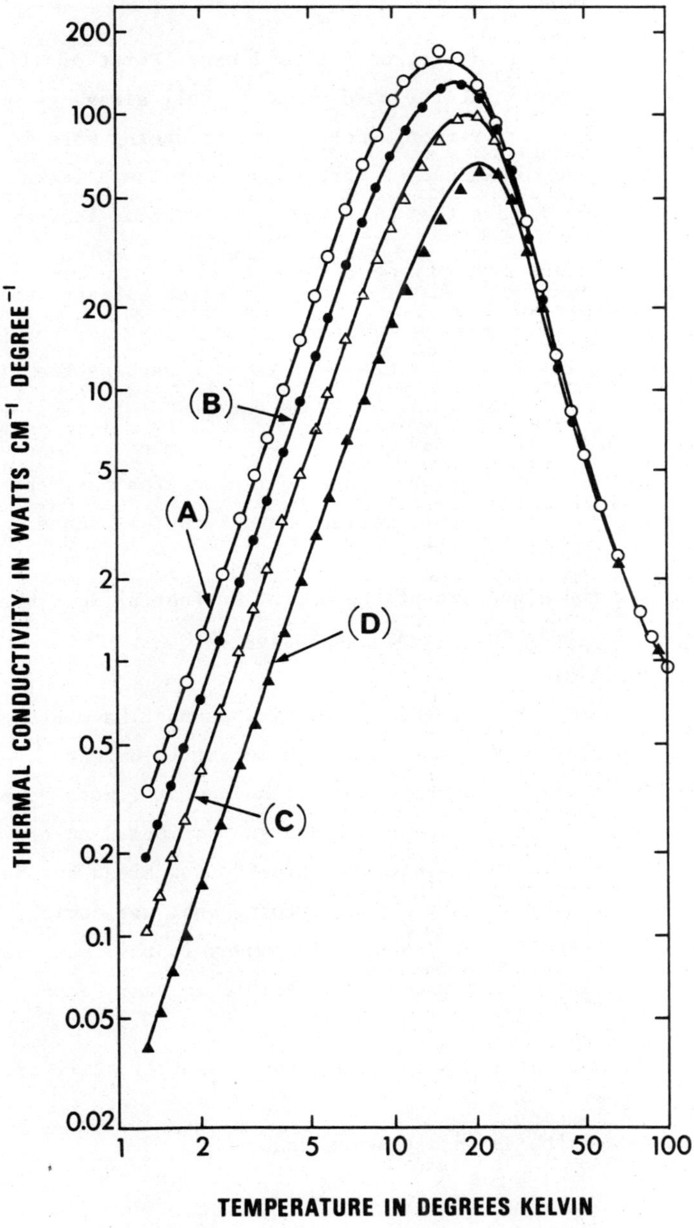

Fig. 3.4. The dependence of heat conductivity on specimen size for lithium fluoride. Mean crystal widths: (A) 7.25 mm, (B) 4.00 mm, (C) 2.14 mm, (D) 1.06 mm.

Discussion

Elliott: Perhaps Professor Peierls would like to begin by clearing up this little difficulty about the p's.

Peierls: I will be glad to do so, as far as I can. First of all, it is true that from the construction of the word - and I shall always go on regretting that I am responsible for bringing such an ugly-sounding word into the literature - there should indeed be three p's. But there is an obscure rule in German grammar which says that when your construction leads you to a triple letter, then you leave out one and you end up with two only, so two p's (or p^2) is the correct version. What Dr. Berman's slide illustrated was simply that in 1929 I was a very poor proof-reader, and I have not changed in that respect - I could produce other evidence! This is perhaps the right context in which to remark that, though it would be ungrateful to cavil at the poetic licence with which Dr. Berman has glossed over a number of major errors, and given the impression that they were just sort of confusions, the one he pointed out was actually rather a big howler - and I could produce a list of others!

Elliott: What was the signature at the bottom of that slide - what is the surname of the man whose Christian name is log 2?

Berman: Sir Francis Simon.

Kurti: Just one other little anecdote, about what must have been a reprint of a later paper, which you sent to Simon from Zurich to Germany, round about 1932. It was about size effects, I think, and how at low temperatures size effects in conductivity could be found, and you scribbled on top of it - in German - "Is it possible to measure something?" Now Simon by that time was so preoccupied with Nazi Germany and everything that was coming, that he thought that what Peierls really wanted to convey in this very subtle fashion, was whether experimental work was still possible in Nazi Germany - or rather, pre-Nazi Germany!

Elliott: I wonder whether it is the same apparatus they still use as the hydrogen liquifier?

Berman: It was not quite as versatile as that!

4

Disordered Systems

S.F. EDWARDS F.R.S.

I was asked to talk either on a subject which stemmed from Professor Peierls' interaction with me, or on science policy. My talk will be about some aspects of disordered systems in general - not just the particular case of science policy. In order to prepare the ground for a discussion of conductivity in disordered systems, which is the subject Rudi introduced me to many years ago, I'd first like to talk a little bit about phase transitions in disordered systems. This work stems from some discussion I had with Phil Anderson on the peculiar properties which are now being discovered about disordered spin lattices, or spin glasses. If you dissolve an element, which has a spin, like manganese, in another element such as copper, the interaction between the spins is almost a random function, because it oscillates in distance. Since the manganese atoms are at any old place in the lattice, the interaction energy is sometimes positive, sometimes negative. Yet it is quite clear experimentally that this thing shows a phase change. For example, Fig. 4.1 shows the susceptibility of such a solution as a function of the temperature T. The system is in complete chaos, but there is clearly a phase transition there.

Now the one thing I did learn from Professor Peierls is that physics is all one, and if you are ever in a jam in one particular area you should just think about some other area, and you will probably find somebody has already solved your problem there. A completely disordered system that mathematically is very similar to this one concerning magnetic systems is the problem of what happens when you try to cook an egg. An egg consists of very long molecules,

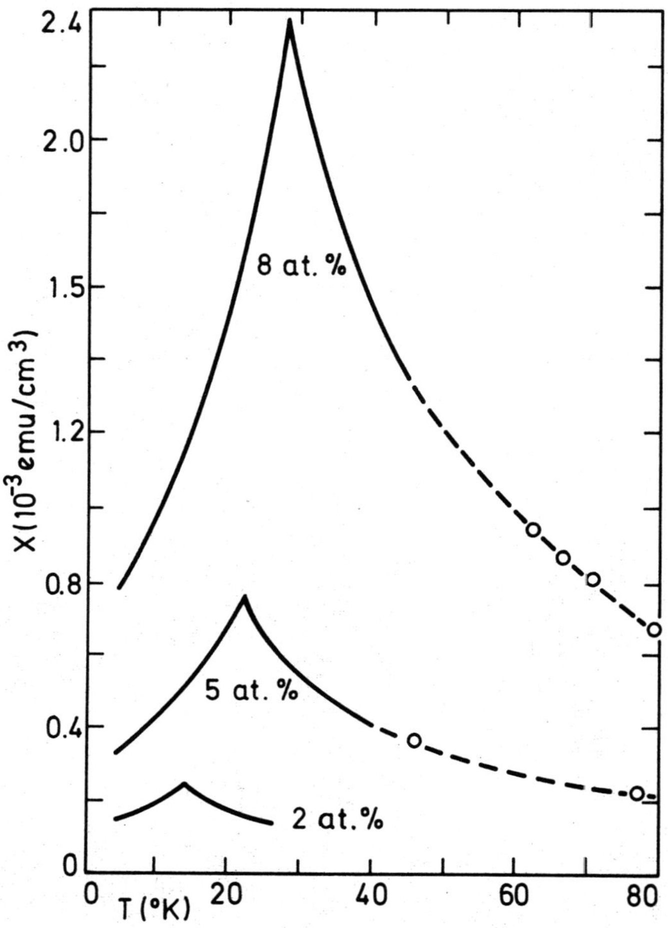

Fig. 4.1 Magnetic susceptibility X of a solution of Fe in Au as a function of temperature at various concentrations. (From Cannella & Mydosh, Phys. Rev. 6B, 11 (1972) 4220).

and as you start warming it up and thus improving the interactions, although it is completely chaotic there comes a point where the thing changes essentially from a liquid to a solid, with apparently no great change within it. How can one describe a system where molecules are able to wander around, yet somehow the system has made a transition from a liquid to a jelly? The answer is, look at one of the molecules and keep on looking at it, and if it stays in roughly the same place you are in a solid, while if it moves around then you are in a liquid.

Let's try and apply that kind of idea to a spin glass. We have a collection of randomly-located spins; let us call $\underline{n}_i$ the spin which is located at the point i. Then there is an interaction between the spins of the form

$$J_{ij} \; \underset{\sim}{n}_i \cdot \underset{\sim}{n}_j \tag{4.1}$$

The associated Boltzmann factor is

$$\exp -\Sigma \, J_{ij} \; \underset{\sim}{n}_i \cdot \underset{\sim}{n}_j / kT \tag{4.2}$$

J_{ij} is known to be oscillating, and can be regarded as a random variable. Now, suppose I look at spin $\underset{\sim}{n}_i$ at one particular time 1, $\underset{\sim}{n}_i(1)$, and at another particular time 2, $\underset{\sim}{n}_i(2)$, and evaluate $<\underset{\sim}{n}_i(1).\underset{\sim}{n}_i(2)>$. If the spins are pointing in a definite direction, albeit chaotic, you expect this quantity to be non-vanishing, but if the spins are having the normal thermal excitations, they can point in any old way and it will be zero. In fact, any simple calculation of a disordered system of this kind will always give you zero for $<\underset{\sim}{n}_i(1).\underset{\sim}{n}_i(2)>$.

To investigate whether $\underset{\sim}{n}_i(1).\underset{\sim}{n}_i(2)$ can have a non-zero value - let us call it q - we follow a similar argument to that used in the Curie-Weiss theory of ferromagnetism. Using the fact that J can be regarded as a random variable, the effective joint probability for finding the spin at i at time 1 and at time 2 involves the factor

$$\exp \left\{ (\underset{\sim}{n}_i(1).\underset{\sim}{n}_j(1)).(\underset{\sim}{n}_i(2).\underset{\sim}{n}_j(2)).\rho/(kT)^2 \right\} \tag{4.3}$$

where ρ is the mean square value of J_{ij}. If the variables $\underset{\sim}{n}_i$ and $\underset{\sim}{n}_j$ are completely random as far as the sites are concerned, one can derive the approximate equation

$$q = \coth\,(q\rho/k^2T^2) - \frac{k^2T^2}{q\rho} \tag{4.4}$$

We all know what this kind of equation says: at a certain critical temperature q stops being zero and has a value. In other words, at a certain temperature these spins will settle into some direction, and that direction just depends on the local field around them. To the external observer, this direction may appear to be completely random, but those spins know they have to point in that particular way and you get a phase change.

Now let's try and make this more rigorous. We have to work out the free energy of the system, which will be a function of the quantities J_{ij}; I shall write the whole set of the J's as simply J. Now the particular configurations of the manganese atoms are fixed, so the mean free energy of the system, as measured thermodynamically, is given by

$$F = -kT\ln\int \exp\left(\frac{\Sigma J \underset{\sim}{n}\underset{\sim}{n}}{kT}\right) \Pi(d\underset{\sim}{n}_i) \tag{4.5}$$

If the sites of spins were allowed to move around, one could average in the exponent. The quantity (eq. 4.5) is a difficult thing to evaluate as it stands, but I shall make the assumption that the integers form a continuum, and once you do that the calculation becomes quite easy. Consider the calculation in Eq. 4.1 to Eq. 4.4 which involved quantities of the form

$$\underset{\sim}{n}_i(\alpha).\underset{\sim}{n}_j(\alpha)\ J_{ij}/kT;\ \alpha = 1,2 \tag{4.6}$$

Let's repeat it for a number of values of α, where α runs from 1 to N (and we shall let $N \to 0$ eventually), and evaluate

$$\langle \exp\ (\sum_{\alpha=1}^{N} J_{ij} n_i(\alpha).n_j(\alpha)/kT)\rangle \tag{4.7}$$

This is analogous to the statistical mechanics problem of a system in the same configuration in each of N boxes, and the result is that, in the limit as $N \to 0$, the average (Eq. 4.7) equals

$$\underset{N\to 0}{\text{Lt}}\ \exp(-N\langle F\rangle/kT) \tag{4.8}$$

where F is the true free energy of the system.

Now this may seem to be a peculiar way of working: we think of N boxes, we solve the problem because N is an integer, and then we let $N \to 0$! But let me assure you it works and it gives some very pretty answers. It gives a susceptibility curve that looks exactly like Fig. 4.1. It also gives a value for the specific heat of the system which is proportional to the temperature as the temperature tends to zero, which is a very peculiar thing; it had been foreshadowed by Walter Marshall some years ago, but it is not an easy thing to derive. Anyway, this argument shows that one can get a phase change in the system which is apparently totally disordered both before and after.

Let me give you another example of this same argument, working more towards the conductivity problem. This example is obtained if you look at a vanadium oxide, $Na_3V_2O_5$. This substance is very complicated and I will simplify the description of it for the purposes of this discussion. What happens with this stuff is that at a low temperature it has a very low conductivity, but at a certain temperature it starts conducting in good style, and the explanation which can be offered is as follows. In the solid, the sodium complexes are at essentially random sites. These sites can carry an electron, and at T = 0 they will all carry electrons. But once the system gets hot, the electrons start floating away, and at this point the screening description of the system completely alters. So, what one is saying in effect is that this is a totally disordered system, but at a high temperature you'd simply think of it as electrons floating around with ions present, and the interaction between an electron and an ion is the screened Coulomb potential. But once the system begins to get cold, once electrons start to settle, the screening fails and the electron then sees a straight Coulombic force which is sufficient to bind. So what you find in a system like this is that it has zero conductivity, or the conductivity appropriate to a very poor conductor, until a certain critical temperature and then the electrons get free, and it starts conducting. So you get, again, a phase change in a system which is completely disordered, inasmuch as the ions are on no particular sites in general, and, nevertheless, it is a highly cooperative phenomenon among the different centres.

With these examples in mind, I'd like to talk a bit now about the problem of electrical conductivity in disordered systems. This is a problem which has been on the stocks now for twenty years or so, and all sorts of new twists and discoveries are coming in which range all the way from very theoretical discussions down to problems which are quite practical and to do with semiconductor devices. The kind of difficulty you find is this. If you take an ordinary solid where there is a random potential present - shall we say, if you took a liquid metal - then in the simplest theory of a liquid metal you say everything is dominated by a Schrodinger equation of the form

$$(E - K - V)\,\psi = 0 \qquad (4.9)$$

where the quantity V which appears here is something with hills and dales, some random variables. Then you can ask two questions, amongst others. What is the density of states n(E) which you get out of this problem? And: what is the conductivity $\sigma(E)$ which you get out of this problem?

As regards the density of states, you can make some guesses as to what you'd expect. You know that at a very high value of E, the electrons will probably rise right above the potential so you would expect to get a density of states which is the free electron value, going like the square root of the energy. If you had nothing but free electrons you would approach zero like $\sqrt{E}$. If, however, you have a potential which has both hills and valleys, there is a possibility of finding a very deep hole, and in a very deep hole the electron obviously can be trapped and have a bound state. It's not obvious that such a bound state will remain bound when you realise there will be similar holes at other places in the solid, but it is generally believed that true bound states can exist in this system, and that means that there will be some remote probability of finding very large negative energies. You would then expect the true behaviour of n(E) to look something like that shown in Fig. 4.2.

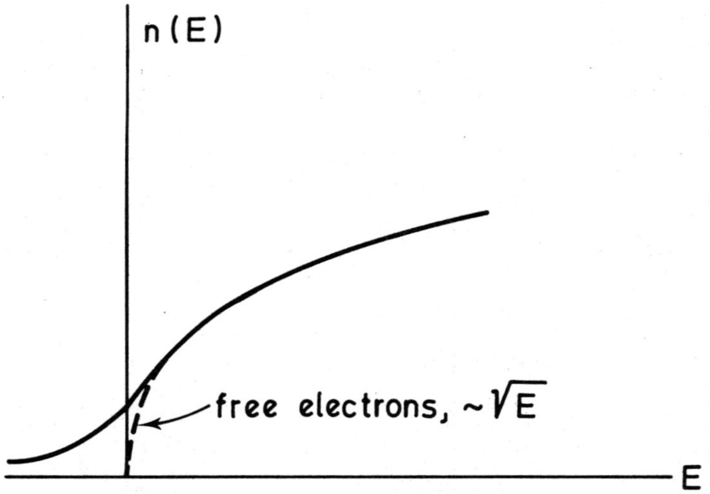

Fig. 4.2. Electron density of states function n(E) in a liquid metal.

There is general agreement about the high E and low E parts of the curve, but the intermediate part is rather difficult to calculate. Let's ask what will happen to the conductivity of such a system as a function of E, $\sigma(E)$. Now it becomes much more difficult, and the general consensus of belief is that the conductivity will have a structure something like that shown in Fig. 4.3. At high values of E the conductivity will just take on the free electron value, it will be very large, and it will follow n(E), roughly speaking. But once you get down to the very low energies, you see states which are trapped in holes in the system, deep potential wells which by some fluctuation occurred, and you would expect that these will not contribute to conductivity. It is, of course, an arguable point whether $\sigma(E)$ ought to tail off below E = 0, as I have drawn it, or whether, as Mott has suggested, it should suddenly drop to zero for E > 0, as indicated by the dotted line in

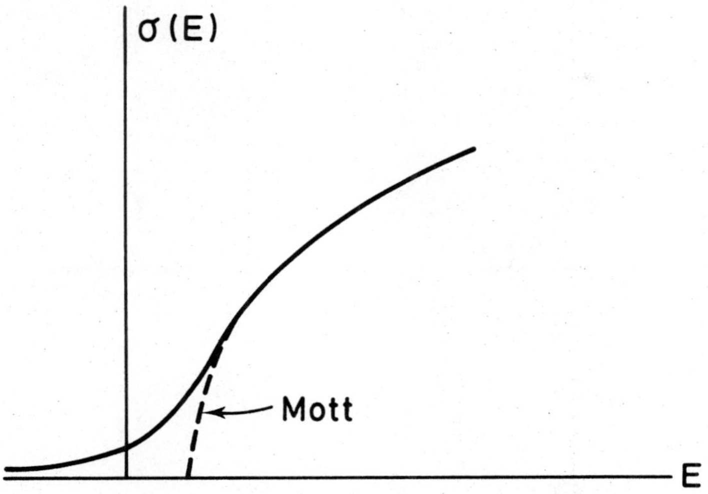

Fig. 4.3. Possible dependences on E of the conductivity of a liquid metal.

Fig. 4.3. I won't go into the arguments for the sudden drop, as they are of a fairly technical nature, which I don't think will affect the argument I am putting forward.

How can one set up a system in which true continuum states exist below where the conductivity has gone to zero? If you attempt to do this in any conventional kind of way, you seem to run into most terrible difficulties, but I think I have recently seen a way how this can be done formally. The method I shall use to describe electrons in a disordered system is that of second quantisation. Now, for the purpose of this lecture I won't describe in detail this method. I would assume that this audience is well aware of it anyway, but I will just remind you of what it means in terms of diagrams - that is, perturbation theory for the electrons. The kind of diagrams you get are as follows: an electron is represented by a line (Fig. 4.4(a)), and the scattering of an electron by a disordered centre by Fig. 4.4(b). If the disorder is extremely localised, you could contract the wavy line to a point, getting Fig. 4.4(c). A double scattering process in which the two scatterings interfere with one another is represented by Fig. 4.4(d); for localised impurities this collapses to Fig. 4.4(e). In a full second-quantised theory of the electron's interaction with the impurity field (by analogy with the

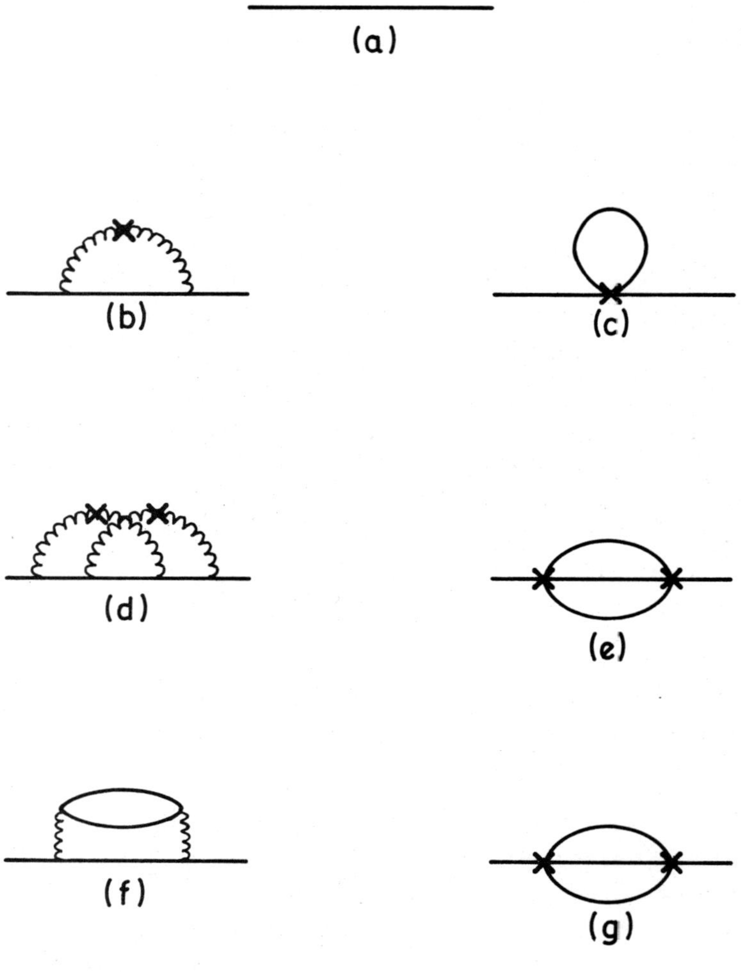

Fig. 4.4. Diagrammatic representation of contributions to the scattering of an electron by impurities (see text).

electromagnetic or pion field cases), there is the possibility of pair creation, represented by Fig. 4.4(f), which can be collapsed, for local impurities, to Fig. 4.4(g). The point I am making here is that if you try to think of the impurities as a field variable instead of as scattering centres, you will inevitably start putting in diagrams like Fig. 4.4(f), which ought not, in fact, to be there. The combinatorial factors will come out wrong, and the answer will be wrong. That has always seemed to be a very severe drawback to using the ideas of second quantisation; one can only really work in first quantisation in such a case.

How can one get around this problem? Physically, all it means is that the disordered field, which represents the scattering centres, is fixed. It doesn't have any dynamics and it doesn't respond to the electrons. If it responded to them, diagrams such as Fig. 4.4(f) would occur and the numerology you get in treating this problem as if it were electrodynamics would be correct. How do we get around this difficulty of terribly overcounting these things? Well, the answer is by exactly the same device as we used on the spin glass problem. Instead of thinking of an electron as having only one wavefunction in its ordinary Hilbert space, we think of a whole set of N-fold Hilbert spaces with electron wavefunctions $\psi^{(\alpha)}$. We write down the effective Lagrangian of this sytem, assuming that in addition to the ordinary propagators of quantum field theory - which I'll now have to write as

$$G = \langle\psi^{(\alpha)}\ \psi^{(\alpha)+}\rangle \tag{4.10}$$

- there is also a possibility of a different kind, which I'll denote by

$$F = \langle\psi^{(\alpha)}\ \psi^{(\beta)+}\rangle \tag{4.11}$$

What does this mean physically? The quantity G tells you things about an electron with an energy E. What the quantity F says is, if I had an electron today at an energy E and then I had another electron tomorrow at the energy E, am I likely to find them in the same place? Are they liable to be correlated with one another? If you put it that way, it's pretty obvious where they would be, because if the electrons are free you would expect F always to be zero, but if the electrons were trapped in some bound state you'd expect F to be non-zero. That's really the only comment you have to make. Let's consider N fermion fields which are interacting with the field of the disordered potentials, which we can now average away and treat as if it's a dynamical field. So I just have N fermion fields all interacting with one another. So it's an N-fold Hilbert space and I can work out what F and G come out to be.

Let us write

$$a = \mathrm{Im}\ (G + F) \tag{4.12}$$

and

$$b = \mathrm{Im}\ (G - F) \tag{4.13}$$

Then it turns out, after some mathematics which I'll simply assume, that the density of states, n(E), is given by

$$n(E) = \mathrm{Tr}\ (a + b) \tag{4.14}$$

and the conductivity, $\sigma(E)$, is given by

$$\sigma(E) = \mathrm{Tr}\ (\nabla \tilde{a} . \nabla \tilde{b}) \tag{4.15}$$

The derivation is lengthy and I can't attempt to justify it here. What is remarkable about the result is that it can lead to a zero conductivity. This is a difficult thing to arrive at, because whenever one tries to calculate a conductivity it is persistently positive - one can't really get a negative conductivity. It is difficult to make it equal to zero, because the slightest approximation will always give you a positive alue for a thing which is manifestly a square of something. The only way you will get an honest zero, in fact, is to have the thing that goes into the integral (in the Trace) equal to zero. Now the phase change that seems to take in this particular problem is that the quantity b in Eq. 4.13 actually vanishes below a certain critical energy, so that from Eq. 4.15 the conductivity vanishes at a certain energy. The density of states, on the other hand, does not vanish. Once you get above the transition then a and b turn out to be the same thing, and the conductivity is non-vanishing.

The variable which I have called a, is like the q of the spin glass. We have a phenomenon which I might call screening of an electron glass; it is analogous to the situation of the electrons in vanadium oxide - they either stick or they don't stick. So now one can attempt to do a calculation. It must have been 18 or 19 years ago when Professor Peierls set me the problem of trying to see if one could get the conductivity out of the set of diagrams I've been talking about - not, of course, including Fig. 4.4(f), which, as I've pointed out, is properly absent. One can show that one can deduce Eq. 4.15 - this formula was actually in Dr. Greenwood's thesis. He is here today, and he stimulated the whole of this development. It was not written then in quite this form, but in an equivalent form, because above the critical energy $(\mathrm{Tr}(\nabla \tilde{a} . \nabla \tilde{b}))$ becomes simply $\int (\nabla \rho)^2$, where ρ is the density matrix.

By looking at this set of diagrams you can also produce a solution to the Boltzmann equation - not perhaps a great triumph, since I suppose the Boltzmann equation for electrons was sorted out at about the same time as the Umklapp processes. (Certainly the only clear account of it that I can understand is in Peierls' book on the Quantum Theory of Solids). This is a perfectly standard result in standard circumstances; the difficulty is, can one get oneself into the extremely non-standard circumstances of the other kind of problem?

It turns out that it is quite possible to do this. Let us put in some argument for the $\psi^{(\alpha)}$'s:

$$G(\underset{\sim}{r}, \underset{\sim}{r}', E) = \langle\psi^{(\alpha)}(\underset{\sim}{r})\psi^{(\alpha)+}(\underset{\sim}{r}')\rangle \tag{4.16}$$

and likewise for $F(\underset{\sim}{r}, \underset{\sim}{r}', E)$. Now if you calculate G under the elementary circumstances of almost free electrons, it will turn out to be a function of $\underset{\sim}{r} - \underset{\sim}{r}'$, and a and b will be diagonal. But when you write down the coupled equations for G and F and assumed that $G(\underset{\sim}{r}, \underset{\sim}{r}, E)$, $F(\underset{\sim}{r}, \underset{\sim}{r}, E)$ are non-trivial functions of $\underset{\sim}{r}$, then there can be solutions in which this is indeed the case. In other words, if I put $\underset{\sim}{r} = \underset{\sim}{r}'$ in the primitive case I get a constant, but in general I get a function of $\underset{\sim}{r}$. And that turns out to be the key to the whole thing. If we look at the actual values, it appears that the density of states is like $\sqrt{E}$ for large E, and for small E it is like $\exp(-\sqrt{E}/\lambda)$, where λ is a parameter designating the intensity of scattering. In between these regimes there comes a critical energy E_C around which, on solving the self-consistent field equations, you find that fractional powers appear in all the quantities. This is a familiar thing in statistical mechanics nowadays, of course. In this case one finds as E tends to zero that

$$G(\underset{\sim}{r}, \underset{\sim}{r}) \propto \frac{1}{r^{4/3}} \tag{4.17}$$

The conductivity can then be worked out and looks much like Fig. 4.3.

In summary then, it may well be that by putting together a number of different threads one can see the emergence of a dynamical phase change of quite a different nature to the conventional ones which occur in ordinary equilibrium statistical mechanics. You have to be prepared to use quantum field theory in the second quantisation formalism, and you've got to be prepared to use the modern description of transport theory in terms of these rather abstract concepts which arise in the solution to the inhomogeneous Schrodinger equation. I think there are obviously a lot more applications of this kind of method that one has to look at, because the solid state which I've put into this talk, as is all too obvious, is of an extremely rudimentary nature. My argument has been of a highly theoretical type, but in fact you can make all sorts of interesting experimental checks on this kind of thing, particularly with layer compounds and things of that kind.

Discussion

Castillejo: Is there any connection between your approach and that of the renormalisation group?

Edwards: The obvious answer, from my hesitation, is that I do not know. The difficulty, I think, is that I am working in an N-fold Hilbert space and I think the problem which arises here is not so much the renormalisation group as such, but whether you can extend the whole idea of the renormalisation group into an N-fold Hilbert space. I do not know whether you can. Certainly the kind of expressions I get - fractional powers and what not - all look like the kind of thing you expect in that approach.

Thouless: I think I have a partial answer to that. I think that the parameter N that you use is the same as the dimensionality of the field which is used by the renormalisation group people in statistical mechanics, and indeed Ma has pointed out that in the limit $N \to 0$ you get the electron in a disordered system; but the difficulties in applying the method, I think, are quite considerable, and, as far as I know, nobody has made any progress using these ideas in this particular problem.

Edwards: I could perhaps reply that it does seem to work all right; the difficulty, I think, is that people often object to one's dealing with dimensionality as a continuous variable, but there's no difficulty in actually doing the sums.

Matthews: You say you put a lot of well-known ingredients together. It seems to me the crucial thing that you yourself have put in is a rather ingenious, perhaps rather singular, identity. Could you write it in its simplest form?

Edwards: It all starts with the identity (true for $N \approx 0$) $X^N = 1 + N \log (X) + \ldots$ Now, I want in the end to take the average of the F given in Eq. 4.5. This means the average, with whatever the weight factor of the J's is, of the logarithm of the integral involving the exponential of the J's. And here you see the essential problem of the disorder. You cannot average over the disorder when it is in the exponential. You have got to work out the thermodynamics of the system and then average over the disorder. And when you do that you need to integrate that logarithm. But you've got to alter the order of integration, in order to be able to do the calculation. In other words, I can do the integral over J much more easily than I can do the integral over the directions of the spins. So the way around it is to say, using the previous identity, that log X is the coefficient of N in the expansion, for small N, of X^N. We then replace the logarithm in Eq. 4.5 by the Nth power of the integral. This in turn is formally the same as considering N similar systems, giving each $\underline{n}$ a label $\underline{n}^{(\alpha)}$, and inserting $\prod_{\alpha=1}^{N}$ in the integral. Now the order of the J and $\underline{n}$ integrations can be reversed, and standard techniques applied to do the average. When the average is done, the coefficient of N will be the term I want. What it amounts to is, that it is actually easier to integrate over N Hilbert spaces where N is an abstract parameter which can take any value, than to attempt to integrate over the spins first, carrying all the information about the disorder, and then trying to integrate over the disorder in the end. It is the same kind of philosophy that leads one to use the density matrix. It is simply that you like to average over states before you average over energy.

Kurti: May I ask a very simple-minded question? I really only do so because I find it difficult to resist the temptation to receive, publicly, an elementary tutorial in physics from the Chairman of the Science Research Council! But can I go in two stages: first of all, when you talked about spin glasses, I take it the glass is in inverted commas?

Edwards: Yes, indeed, it just means "disordered system".

Kurti: That is all right - we know there exist spin glasses; for example, all the wonderful glasses in church windows are all spin glasses, that is why they are so beautifully coloured. And now we come to the second point: I should like to ask, what is the fundamentally new approach as compared with the old-fashioned straightforward Weiss type of approach? Because, after all, one has known for a very long time that if you take a nickel-copper alloy, depending on the concentration, it turns ferromagnetic at different temperatures.

Edwards: Well, I think there are probably people in this audience who are more expert in the history of this subject than I am, but the thing I think that produces a special fascination is the following: it is quite easy to produce a rough theory of ferromagnetism by invoking the idea of clusters - sooner or later you can come out with a mean field which is non-zero. The surprising thing about what I have been describing is that there is never a mean field at all! Explanations have been offered for the behaviour of the spin glass in terms of domains, but the point is that there does not seem to be any reason at all why there should be such domains. The thing really is a solid solution. And yet, even though there is complete disorder, with every parameter in sight averaged over, there is still a phase change. I think it was the fact that the experiments showed the susceptibility cusp on a material in which there was no trace of any mean fields at all that aroused all the interest in this subject. Of course, by putting on an external field, the cusp is destroyed - it rounds off, as you would expect.

Elliott: You made the analogy earlier between a disordered solid and an egg, but I did wonder at the time if it was not perhaps in some ways more like the curate's egg, in that it differed in different parts. There have been suggestions that in a really disordered material you would not get a well-defined phase transition, because there was always a finite probability of finding somewhere around a macroscopic piece of perfect crystal, which would therefore have a transition at one temperature. So in the system as a whole, you would have a gradation of transition temperatures rather than an exact one. Now, it does appear that in disordered systems, in some of them certainly, there are well-defined transition temperatures - the spin glass is one. And I wondered if you had any comment as to whether there was any fallacy in the earlier argument, or whether there were conditions under which you would expect a well-defined transition temperature or not?

Edwards: Well, if you have got a solid solution, say manganese in copper, and you quench it rapidly, then it is reasonable to suppose that the distribution of the manganese ions is exactly the same as it would have in the liquid. In other words, you can calculate the chance of it being a perfect crystal. And it just is not there. Now that is very different from the situation in which, shall we say, you quench polythene. This substance furiously crystallizes at the slightest chance, and so if you get a piece of so-called amorphous polythene, it is always full of little crystals, and that is a terribly difficult problem to treat. But I do not think that would occur in this particular substance, and you could arrange that it could not. If I can go back to the polymer field, it is certainly possible to get polymers in which the constituent monomers are synthesized at random. Such things cannot crystallize. They must be amorphous, and there is no question of any domains ever building up in them. So I think the answer is, if we go back to the solid solutions, that it is also true that, if you wait long enough, your metals will always have phase separation, in general circumstances. So, I think, it is possible to have a perfectly amorphous material which is as disordered as the laws of physics permit, and that still shows a phase change.

5

Some Applications of Kapur -Peierls Resonance Theory

G. E. BROWN

There are a couple of people here who might more appropriately talk about the Kapur-Peierls theory (1): a long time ago Tony Lane wrote a long manuscript proving the completeness of the eigenfunctions. I think he lost this manuscript - certainly it was never published. Also, the more interesting recent developments, which I'd like to talk about later were done more by Lane and Robson than by me. However, some people have been asked to speak because they had something to say, some people have been asked to speak because they are great men, in England at least, and I've been asked to speak because I had the sheer persistence to spend ten years in Birmingham, I suppose. In any case, probably many people don't know what the Kapur-Peierls theory is, so I thought I'd begin with a short introduction.

I'm not really going to go into the many-body problem very much because it's complicated and one can understand the main features of the Kapur-Peierls theory by talking about the one-body problem. I think the Schrodinger equation is quite familiar to people. Inside a radius R, one has the Schrodinger equation with some potential; outside R one has the free field equation,

$$(E - H)\phi = \frac{\hbar^2}{2m}\frac{d^2\phi}{dr^2} + \left[E - V\right]\phi = 0, \ r < R$$

$$\frac{d^2\phi}{dr^2} + k^2\phi = 0, \ r > R \tag{5.1}$$

and the idea was to solve this equation by expanding in eigenfunctions, ϕ_n which are chosen to be solutions of the Schrodinger equation with a particular boundary condition. (Here I'm neglecting Coulomb waves and all of those things. We're just dealing with S-states).

$$\frac{\hbar^2}{2m}\frac{d^2\phi_n}{dr^2} + \left[W_n - V(r)\right]\phi_n = 0 \quad (5.2)$$

$$\frac{d\phi_n}{dr}\bigg|_{r=R} = ik\phi_n\bigg|_{r=R} \quad (5.2.1)$$

The main idea is this boundary condition, something I would call a rather physical boundary condition; that is, the logarithmic derivative of the eigenfunctions is chosen to be that of an outgoing wave; since one wants ultimately to describe problems involving outgoing waves this turns out to be quite convenient for many purposes. Now, the one-body wave function - here I'm following the Kapur-Peierls paper briefly - is expanded as a sum over these eigenfunctions, but of course the problem is that the solution to the Schrodinger equation does not satisfy this boundary condition. Therefore, one cannot expand it just in terms of functions which do satisfy the boundary condition, and Kapur and Peierls had this tricky little function χ, which was added in, which just got this right.

$$\phi = \sum_n a_n\phi_n + \chi. \quad (5.3)$$

In the end this is a function which is let to go to a limit where it has no value but it has a derivative, which is chosen to make the derivative of the wave function continuous at R. This whole business has been done in a more satisfactory way in later years using Green's functions, in which case one doesn't have to deal with the little Peierls demon, χ. In any case, then it's a straightforward matter - aside from this χ - to get the entire solution, the S which is the coefficient of the outgoing wave, and the way that this is done is simply to join the inside and outside wavefunctions and their derivatives at R.

$$\frac{S}{I} = \left\{ \sum_n \frac{\hbar^2}{2m}\frac{\phi_n^2(R)}{(W_n-E)}\frac{e^{-2ikR}}{N_n} - \frac{1}{k}\sin kR\, e^{ikR} \right\} \quad (5.4)$$

Then if one goes on to the many-body problem, the many-body problem goes very much in the same way, except that in addition to the odd particle outside, one has the function $\chi_o(\xi)$ describing the A particles in the core, and their collective coordinates are described by ξ - the 3A coordinates of the core. Then it's quite convenient to take the wavefunction and expand it in terms of a single odd particle - that would be the function $\hat{\psi}_m$, and the eigenfunctions of the core, because then it's quite easy to state the boundary condition - namely the boundary conditions are stated just on this odd particle.

$$\Psi = \left(\frac{\sin kr}{k} + Se^{ikr} \right) \chi_o(\xi), \quad r > R$$

$$\Psi = \Sigma a_p \Phi^{(p)}, \quad r < R \tag{5.5}$$

$$\Phi^{(p)} = \sum_{j,m} a_{j,m}^{(p)} \chi_j(\xi)\hat{\psi}_m(r) \tag{5.5.1}$$

Now in all of these theories only one particle is allowed to get out of the nucleus. As soon as one begins allowing for deuterons and tritons and α particles, one gets into a horrible mess, and every year one or two people come along with papers claiming that they've solved these messes, but soon somebody else comes along and proves that they've not. This is an old problem in nuclear physics, which may one day be solved but certainly has not been solved.

Well, when I went to Birmingham, I'd been quite turned off by nuclear physics - I didn't do that as a thesis before, but I'd been called in on task forces to work on other people's problems, which turns one off even more than working on one's own problem, and I communicated my distaste for nuclear physics to Prof. Peierls, because I was then going forging ahead on quantum electrodynamics (without making much progress, but still forging ahead!); he pointed out that there were lots of interesting problems in nuclear physics. It took me some years to understand the point that there were some interesting problems, but I do remember that he had already, through sum rules, understood what Vicki Weisskopf had been going round the world saying for some years already, with his slides, which were getting so dirty by his thumb prints that one could hardly see the contours. Older people will remember; there were three-dimensional slides where things went up and down in sorts of plateaus and hills, and they were done on a sufficiently broad perspective that you couldn't really see whether experiment and theory agreed very well, but it looked like kind of the same mountain range - I mean the theoretical one was not the Himalayas and the experimental one the Cotswolds. They were in the same dimension. So everybody was trying to understand these points, and Vicki was talking about these giant resonances, which turned out later not to be giant resonances at all, but if I remember correctly, something like the Ramsauer effect - I'm not going to explain what the Ramsauer effect is - anyway, everybody thought these were giant resonances. Professor Peierls had already at this time had a picture where he pointed out that the coefficients, when the core of the nucleus is in the ground state, would tend to peak up (these coefficients are called partial widths) so that one would tend to see the maxima in the cross section at these points, and they peaked up at the points where one has single particle resonances in the average well, which was the optical model well, and which people knew about at this time. The situation is illustrated in Fig. 5.1.

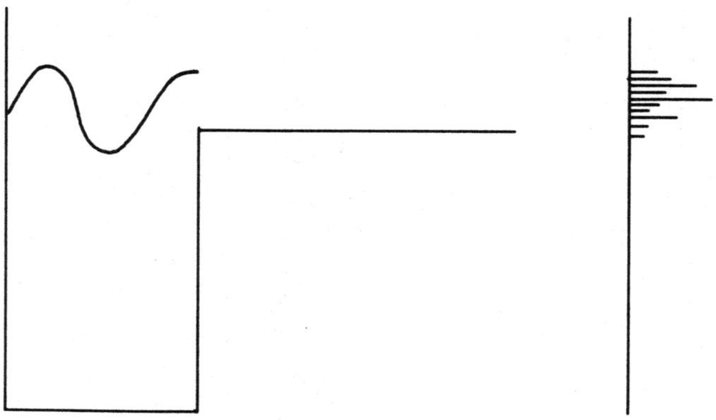

Fig. 5.1 Schematic representation of shape resonances and partial widths.

Now, I had heard about these discussions and they looked fairly convincing, but nobody at Birmingham was allowed to publish anything because the theory had not been antisymmetrised. If one could have treated this odd particle as a π meson, it would have been quite all right, but at that time people weren't scattering π mesons off nuclei. So the picture was really all there, and not really very much different from what it is today, and the antisymmetrisation, I think, has only generally obscured the issue, although it's something that was done later by Bill Friedman at M.I.T. in a satisfactory way for the first time. This picture is now called the Lane, Thomas and Wigner picture, because Lane, Thomas and Wigner independently reached the same picture, and then they met some place and I guess that they were surprised that they all knew the same facts, and at that time people tried to publish one paper rather than three papers. That's not meant to be funny, by the way.

In many cases one already understood from arguments of Feshbach, Porter and Weisskopf that one ought to look at the shape resonances or the giant resonances as coming from the S-matrix that was averaged over energy, and if one averaged over a broad band of energy ΔE, then the corresponding particles would come out in time $\hbar/\Delta E$.

$$S = \langle S\rangle_{Av} + (S - \langle S\rangle_{Av})$$

$$\langle S\rangle_{Av} = \hat{S} \tag{5.6}$$

Now, this is a rather amusing picture, because you have at your control - on averaging over the phase shift - a means of getting the particles out faster and faster: if you average over a larger interval ΔE you can get them to absolutely whip across the nucleus. Then it becomes clear that if you really averaged over a large interval ΔE, you'd get back to the point where Born approximation was pretty good - namely, you could consider collisions first with one particle and then with two particles and then with three, and if it wasn't quite good enough with that ΔE you'd take a larger ΔE and make the particles go over even faster. Now, these would not be all of the particles, because if one averages the amplitude one has, in addition, the fluctuations, and the fluctuations would take longer - this was made very clear in an article by Vicki Weisskopf and another Friedman. The single-particle resonances are called shape-elastic resonances, I'm not quite sure why. So then the shape-elastic particles are supposed to be represented by solutions in a complex well and then there is considerable algebra which one can go through, but which is really, I'm convinced to this day, much more simply gone through in the Kapur-Peierls formalism than in any other formalism (and in nuclear physics it was certainly first gone through in the Kapur-Peierls formalism), and one comes out with an expression for the optical model potential which is the average of the many-body V, the sum of the interactions between the incident particle and all particles in the nucleus - which can be written down $V = V(\underset{\sim}{r}-\underset{\sim}{\xi})$. One has an average potential $\langle\chi_o|V|\chi_o\rangle$ and then an additional potential which gives mainly an imaginary part, but there's also a principal value to it which contributes to the real part, and the point is that the averaging over the energy is carried out by adding a large imaginary part - a part that is large compared with the fine-structure, but hopefully small compared with the gross structure, or else you won't find out anything, adding this to the denominator.

$$\hat{V}(r) = \langle\chi_o|V|\chi_o\rangle + \mathcal{W} = \bar{V} + \mathcal{W} \tag{5.7}$$

$$\mathcal{W} = \sum_{p'} \frac{\langle o|(V - \bar{V})|p'\rangle\langle p'|(V - \bar{V})|o\rangle}{E - W_p' + iI} \tag{5.7.1}$$

This formula has since turned up in every other formalism for the optical model. I should explain what the states $|p'\rangle$ are - they're not quite the same as the states $|p\rangle$ I had before - they are the same but with the χ_o, the incident channel projected out: we'd originally made a mistake in this, and Tony Lane, I think, was the referee of the paper - it was always easy to see which papers he refereed because Harwell typewriters have certain characteristics - and he saved us from making dreadful mistakes. In fact, in those days in this game, one only had to have some idea and start it off, and then Tony would correct it to the stage where one could make a perfectly decent publication out of it. In any case, we realised later that one really has to project out these states - in many-body theory it's called taking out unlinked clusters.

Having this formalism you'd think that one would go ahead and compute the optical model potential, but one needs some model for these states $|p'\rangle$ and that's where we really took issue with other people. People had tried frivolous models. The problem in this game is that everything one does works. One always gets the right answer, within factors, but one gets the right behaviour and everything, and so from agreement with experiment or something like that one cannot tell that one has been doing the right thing - one has to think, and that's hard. Well, thinking about it, if one wants to make models for these states, one can either make models for compound states; or, one knows that these states are essentially on-energy-shell, that is, they come in at energies where the denominator vanishes, for the interval one is averaging over, and one knows that the inelastic scattering on nuclei goes mainly by the excitation of vibrational states, that is, one hardly sees the incoherent states, so the natural model to use was one in which one used vibrational states here and the first calculations were really done many years later - the first calculations were reported (2) at Peierls' 60th birthday celebration by Nicole Vinh Mau and Bruneau, who was working with her, and I wanted to just show some results from this time. Figs. 5.2 and 5.3 sort of show what that game is. The point is that from the expression that I showed you the $\mathcal{U}$ is clearly non-local, which doesn't make life easier in plotting things; and in this particular case the results could be simply plotted because a δ-function interaction was used, and in fact the distortions of the incident and scattered waves were not taken into account, but the thing that is kind of amusing is that ($\underline{x}$ is the relative coordinate and $\underline{X}$ is the centre of mass coordinate) the non-locality and everything which is shown in the upper two graphs is pretty much the same as that which was found empirically in a non-local potential by Perey and Buck, although that potential was an empirical one which is found to fit the data quite well. The intermediate states here were chosen -

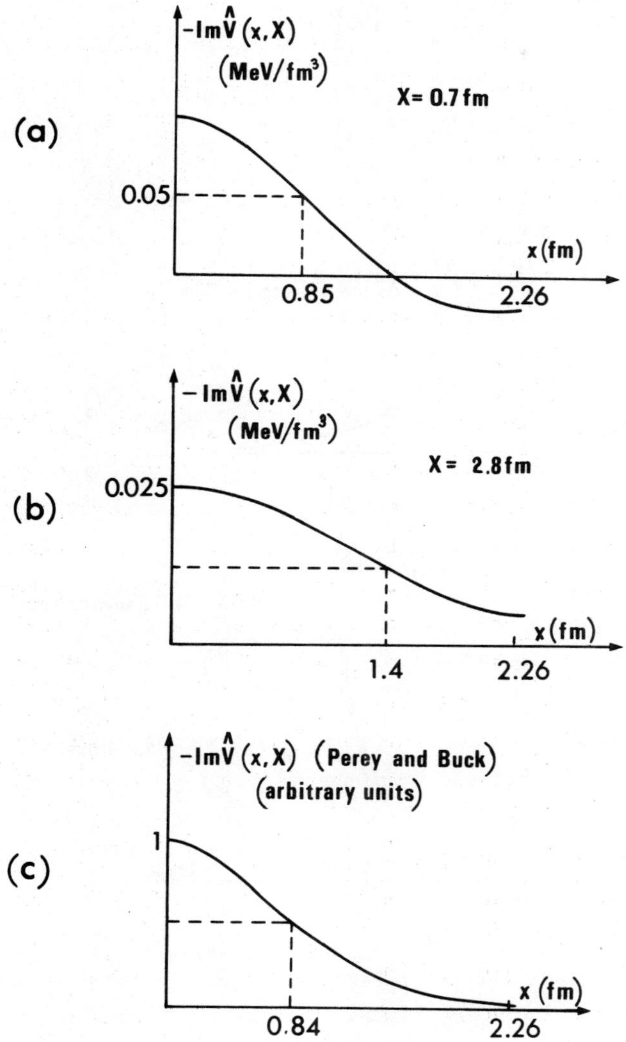

Fig. 5.2. (a) The S-wave part of $\mathrm{Im}\hat{V}(x,X)$ is plotted against x keeping X fixed at 0.7 fm. (b) As for (a), but with X set equal to 2.8 fm. (c) The Perey-Buck non-local potential is drawn for comparison.

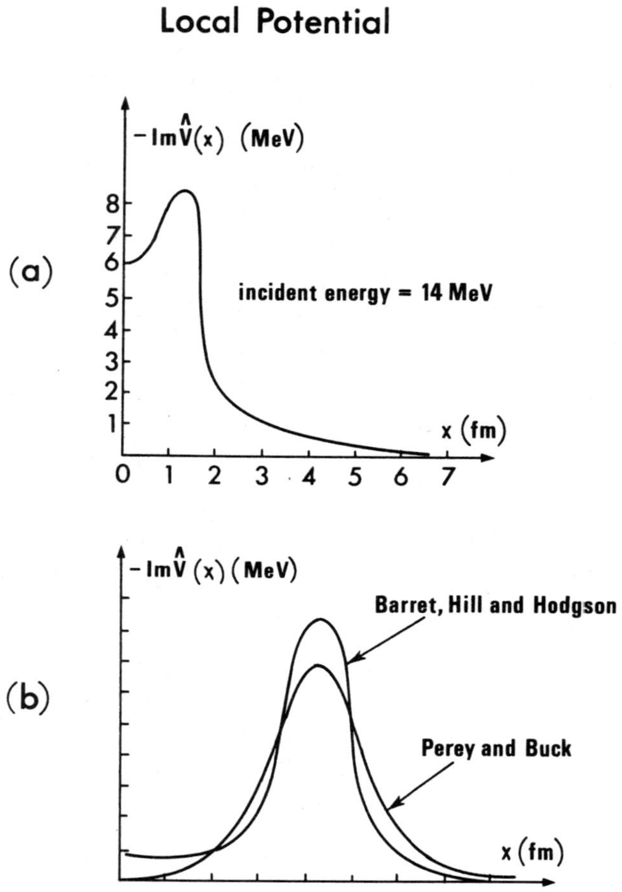

Fig. 5.3. (a) The S-wave part of the Vinh Mau-Bruneau (1) local potential. (b) The local potential curves of Perey and Buck, and of Barrett, Hill and Hodgson, for comparison with (a).

this is for Ca^{40} - to be states consisting of a vibration and a single particle, so it's consistent with the picture that if a particle goes into the nucleus it first excites a vibration. Fig. 5.3 shows what happens when this is converted to a local potential. The amusing thing here is that the local potential which is obtained from the calculation on the top does not look much like the local potential which is obtained from the empirical ones. (The local potentials shown were obtained by integrating over the non-locality coordinate). I believe that now more fancy things have been done, and I have some later results which were presented at the Amsterdam Conference, but this gives the general idea.

What is rather amusing is that in the last two or three years many more people have got into this game - although the theory is roughly fifteen years old, or even older, not many calculations were made until recently, but now a large number of people are getting into the game, and there is predictably nice agreement with experiment, etc. Fig. 5.4 shows some attempts (3) that we made to actually construct these states $|p'\rangle$ by diagonalising in a large space of single particles and vibrations, this is for neutrons incident on Ni^{58}, in other words, states in Ni^{59}; this is actually a model for constructing compound states, a certain subset of them, but those that we think enter into the vibration. The solid line is the theory and the dashed line is experiment, and you can see that there is possibly some sort of correlation between theory and experiment, although it "ain't the greatest", especially if you look at the top, but then this is without averaging over the energy interval. Fig. 5.5 shows what happens if one calculates the optical model potential after averaging over energy. What is amusing is that in the calculations of this model, these giant resonances really do not arise, although of course this is for rather low energy; we'd meant to carry it on up into this region, but Terry O'Dwyer, the man who was doing it, disappeared into University politics. It's also clear from experiments that when one looks in more detail, one does not have these nice shape resonances and single particle resonances which one had thought to have earlier.

Well, although that's taken the majority of my time, it's really neither the main part nor the nicest part, I should say, of my talk. As to the main part of my talk, I think the Kapur-Peierls theory really turned out to be most useful in discussing analogue states. I'd better explain what an analogue state is. Everybody knows about isospin multiplets, and it's just an isospin multiplet if you like. In a nucleus one has a large excess of neutrons, and then by applying a lowering operator to this one will get a state which has the same isospin - this is for a "parent" analogue, you don't need to know all of the terminology - anyway, by using the lowering operator, one gets a state in the nucleus with one more proton and one fewer neutron which has the same isospin symmetry as the parent analogue. This is illustrated in Fig. 5.6. The state obtained in this way will be a fairly excited state in the "daughter" nucleus. What is amazing was that the widths of these states, although they

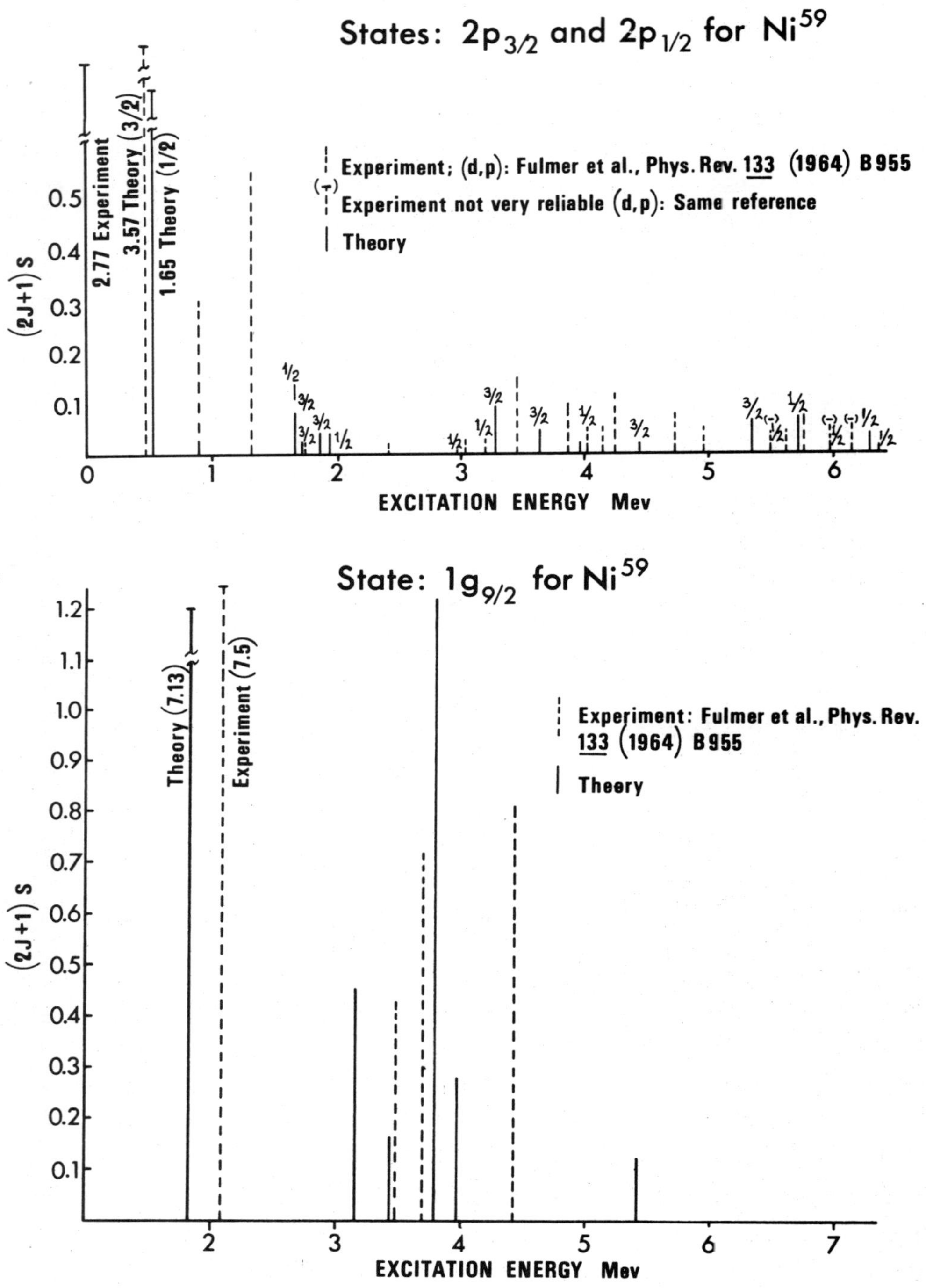
States: 2p3/2 and 2p1/2 for Ni59
Experiment; (d,p): Fulmer et al., Phys. Rev. 133 (1964) B955
Experiment not very reliable (d,p): Same reference
Theory
2.77 Experiment
3.57 Theory (3/2)
1.65 Theory (1/2)
(2J+1) S
EXCITATION ENERGY Mev
State: 1g9/2 for Ni59
Experiment: Fulmer et al., Phys. Rev. 133 (1964) B955
Theory
Theory (7.13)
Experiment (7.5)
(2J+1) S
EXCITATION ENERGY Mev

Fig. 5.4

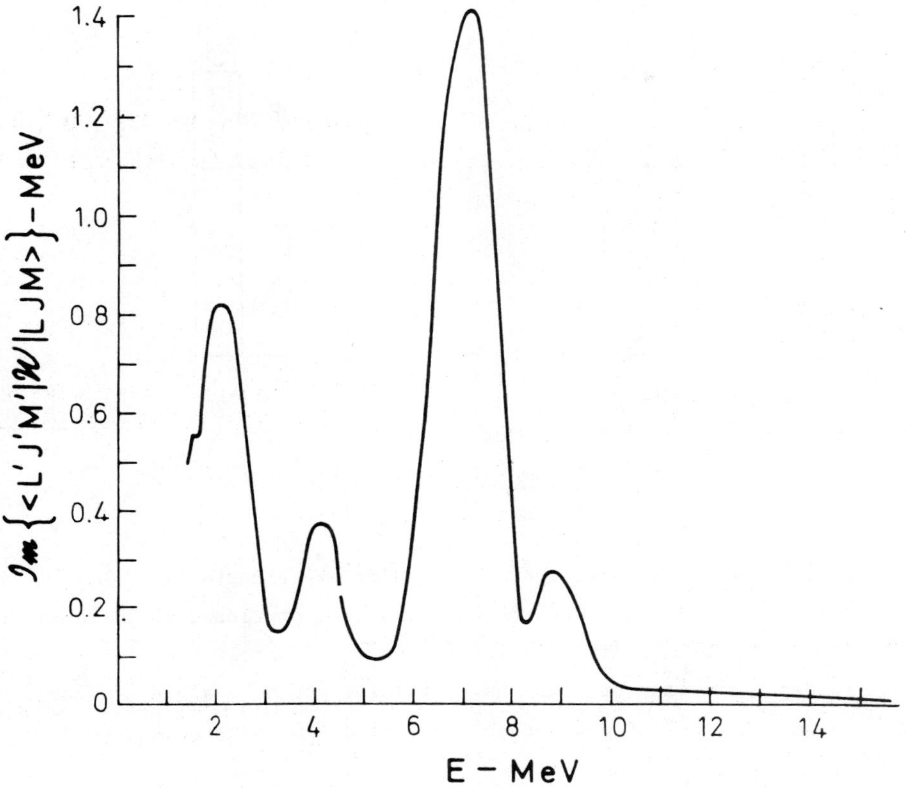

Fig. 5.5. The imaginary part of the optical model potential after averaging over energy.

come at quite a high excitation energy, are rather small, so that they are not just completely dissolved into the background, but one sees them. The analogue has an isospin higher than that of the ground state; nuclei like to have the smallest isospin possible because one knows that the forces are more attractive in a two-body state in isospin-zero than in isospin-one, that is to say, the n-p force is more attractive than the p-p force. So the analogue would be a state at an excitation energy of typically 8 - 10 MeV, and it was quite a surprise to find out that they were so narrow, although Tony Lane and Soper had made calculations which showed that they were quite narrow, but they came out to be even narrower than Lane and Soper had thought. These

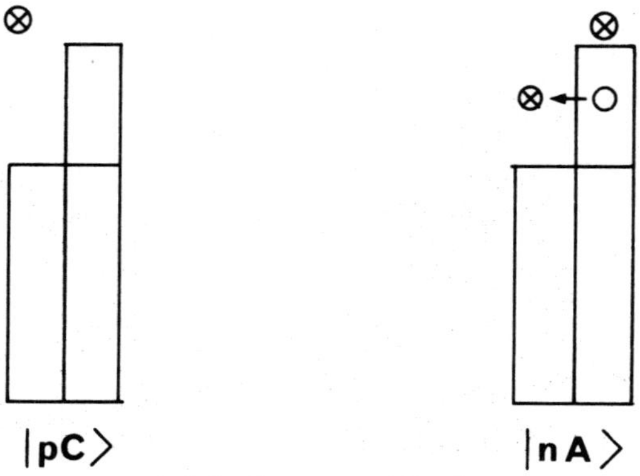

Fig. 5.6. Representation of analogue states.

analogue states have been observed quite a lot experimentally. What is amusing is to think of a nucleus which is really sort of closed shell, just for simplicity, and a proton (protons are over on the left in Fig. 5.6), and a proton is sent into that nucleus; now this will not be a state of good isospin - to make a state of good isospin one needs two pieces, as shown in Fig. 5.7:

$$\Psi_{N-1,\,Z+1} = T_{-}\,\Psi_{N,\,Z}^{P.A.} \;;$$

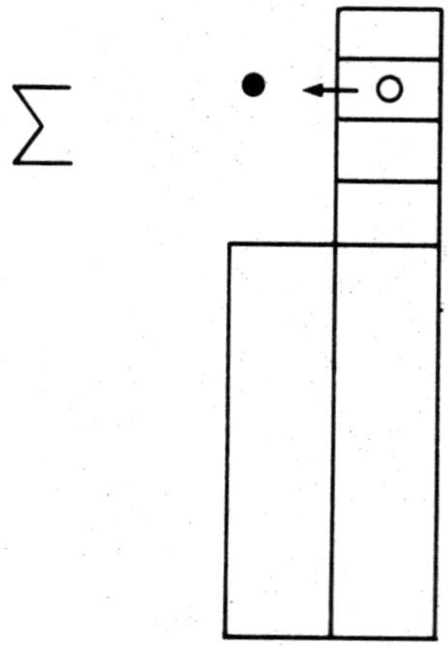

Fig. 5.7.

$$\Psi = \psi_1(r)|pC\rangle + \psi_2(r)|nA\rangle \tag{5.8}$$

I should say before going on that since these states are at a high excitation there will be, often, hundreds or thousands or millions of states, depending on the size of the nucleus, of the isospin of the ground state around in the same neighbourhood. So there's one state of the higher isospin, the analogue state, and then a background of fine structure.

It turns out to be very convenient to treat these first in the Kapur-Peierls formalism, because the p and the n wavefunctions inside the nucleus look pretty much the same. There is a well in which you have the neutron wavefunction, but then the proton wavefunction feels a Coulomb repulsion, so the proton wavefunction will be displaced somewhat upwards, but if you think of integrating out from the charge, the Coulomb potential from a distributed charge is really pretty constant, so the proton wavefunction will look very much the same as the neutron one up to some radius R, so although drawing radii R was always thought to be a tremendous disadvantage in all of these formalisms, and everybody always tried to get away from it, in this particular case it turns out to be a tremendous advantage. One can really talk about a region - people have worked this out in considerable detail - where up to a certain radius R the wavefunctions of n's and p's look very much the same. But then, of course, beyond the radius R they will be different, since the proton can escape, due to the additional Coulomb repulsion in the wells, whereas the neutron is trapped; the neutron is more tightly bound. So it's quite convenient - here I'm still working in sort of the one-body formalism - to use different boundary conditions on the proton: the proton is an outgoing wave, and the neutron would have a boundary condition which is exponentially decreasing. Now one can go through the same game as for the one-body theory in this case, and one gets a deceptively simple result - really it's deceptively simple - namely, in terms of the eigenfunctions ψ_m defined by

$$H|\Psi_m\rangle = E_m|\Psi_m\rangle \tag{5.9}$$

$$|\Psi_m\rangle = \psi_{1m}(r)|pC\rangle + \psi_{2m}(r)|nA\rangle$$

$$\frac{d}{dr}\,\phi_{1m}(r)|_R = ik\,\phi_{1m}(R)$$

$$\frac{d}{dr}\,\phi_{2m}(r)|_R = -\kappa\,\phi_{2m}(R) \tag{5.9.1}$$

one obtains the same formula as Eq. 5.4 with $\phi_n(r)$ there replaced by $\phi_{1m}(R)$.

It's very simple. It doesn't show you any of the complexity of the problem. The point is that the states Ψ_m are not, of course, states of good isospin because although the wavefunctions are the same, or very similar, inside the nucleus, the boundary conditions on the neutron part and on the proton part are different. So it's then convenient to expand these states, which I've called the physical states, in terms of states of good isospin, and this was done by Lane and Robson, and independenly by Mello who was working with me at Princeton. This expansion into states of good isospin is very tricky in that the sums are not absolutely convergent due to the fact that one is expanding functions in terms of other functions which have different boundary values, have different derivatives. But again one can use Green's theorem to do everything - it's always safe to use Green's theorem in this game. If one expands into states of good isospin, one really comes out with a pretty complicated formula, which Robson originally got in the R-matrix formalism. Actually, he got it within the framework of the coupled equations of the Lane theory. I don't think he had, at least not conceptually, properly all of the many-body aspects in, although the form was correct. What comes out is that the S-matrix then is

$$S = \frac{e^{-2ikR}}{2k}\left\{\frac{\Gamma_A}{E_A - E - \frac{1}{2}i\Gamma_A} + \frac{(E_\lambda - E)^2}{(E_A - E - \frac{1}{2}i\Gamma_A)^2} \times \right.$$
$$\left. \times \frac{R_<}{\left[1 - \dfrac{(E_\lambda - E)(\kappa + ik)R_<}{E_A - E - \frac{1}{2}i\Gamma_A}\right]} - \frac{e^{-ikR}\sin kR}{k}\right\} \tag{5.10}$$

with one piece labelled by A referring to the one analogue state, the T_{upper} resonances, and then a whole mass labelled by λ referring to the hundreds or thousands or millions of T_{lower} states. Now this would be pretty much the T_{lower} scattering - the R-function is just the usual sum over widths

$$R_< \equiv \frac{2T}{2T + 1} \sum_\mu \frac{\Gamma_\mu}{E_\mu - E} . \tag{5.10.1}$$

The point that is amazing about this is that due to the interference between T_{upper} and T_{lower} scattering, there is an asymmetry factor on the T_{lower} fine structure that comes out. The E is the unperturbed position of the analogue state, or the upper state, and the E_A is the position it ends up at. There's a sort of Thomas shift, a shift due to the fact that it can decay or, if you like, the principal value contribution shifts it down. So there's an asymmetry factor, shown in Fig. 5.8. The final two Figures show how this business comes out in practice. In Fig. 5.9 you don't seem to see much of

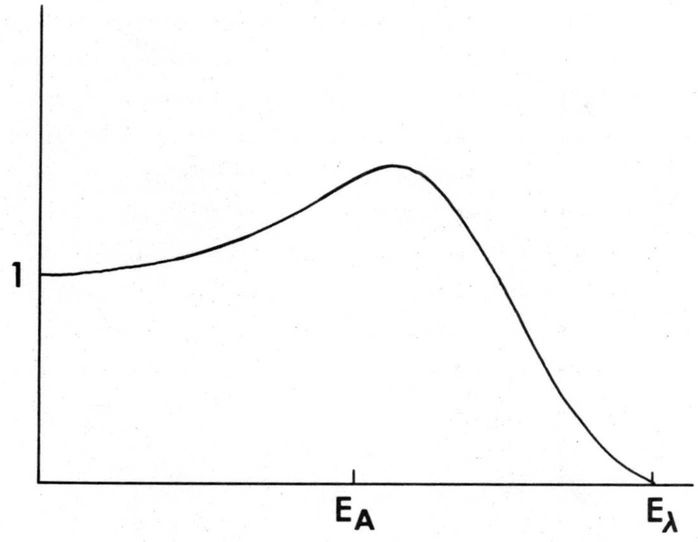

Fig. 5.8. The asymmetry factor $(E_\lambda - E)^2/(E_A - E - \frac{1}{2}i\Gamma_A)^2$.

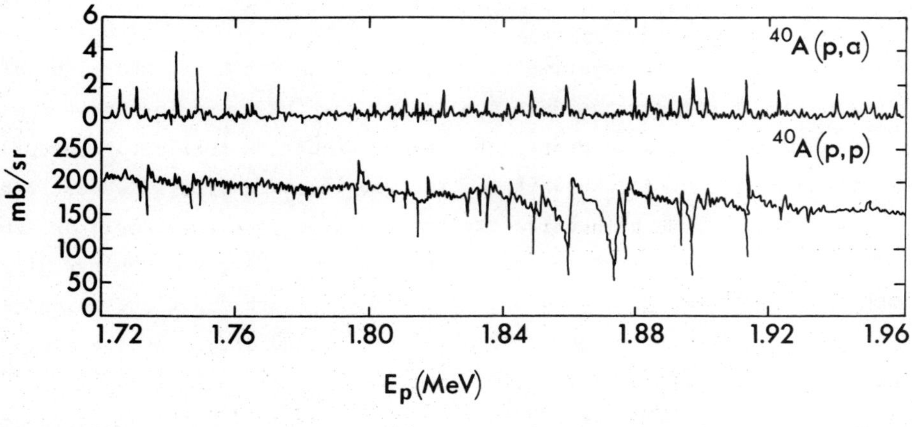

Fig. 5.9.

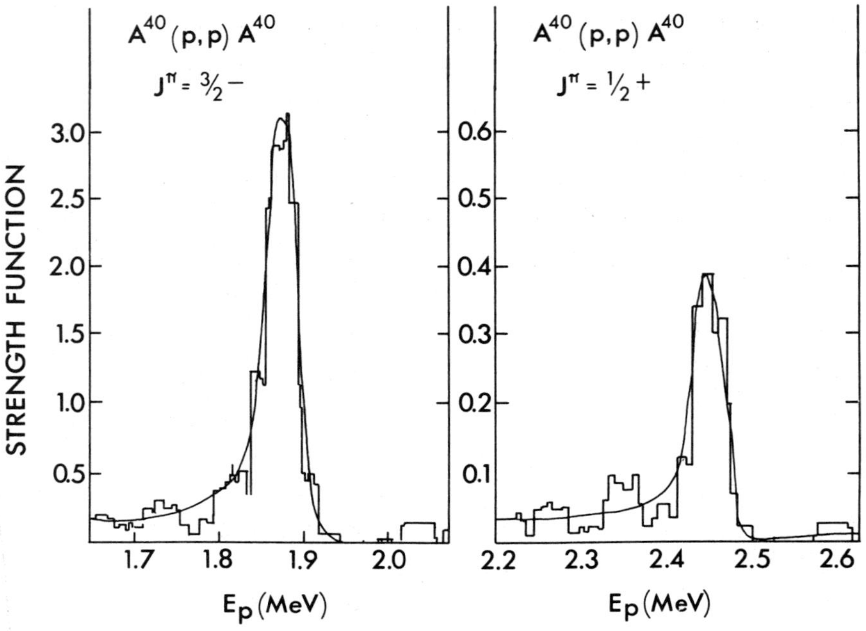

Fig. 5.10. Strength functions.

anything - these are Duke experiments on a small Van der Graaf where, by all kinds of technical tricks, the Duke people got down to a resolution of something like 250 eV - which is pretty good resolution! To solid-state people it may not seem so, but you know if you're sweeping across many KeV's, to have a resolution of a couple of hundred electron volts .... well, people on Keble Road were certainly impressed by this sort of thing. Now, you don't really see much there. But suppose that one plots the strength function, which means that you collect the widths over a 20 KeV interval and divide by the average spacing. That's plotted in Fig. 5.10. And you see the solid line is that asymmetry factor which was plotted in Fig. 5.8, and here the strength function is plotted. Suddenly, when you have this theory to tell you how to plot things, there is this sort of amazing regularity that comes out of the widths.

So I think this is a very pretty interim end to the story, and it's a nice application of the Kapur-Peierls theory which Kapur and Peierls clearly never thought about, but the theory appears to be, really, simply made for this.

Now, I should perhaps add one remark - I really wanted to make a remark at the very beginning - and that was that my first introduction to the Kapur-Peierls theory was that if one does the usual theory as originally printed, there's the hard sphere scattering term. Prof. Peierls always made the point that there must be phase relations between the levels, in that if one changed this sort of arbitrary parameter R, the hard sphere scattering would change, and therefore, these would change. Now, it has been found that these phase relations are precisely those phase relations which Weidenmuller, and also Kerman, McVoy, Kawai and so forth have been worrying about. The way that they come about in the present work of Weidenmuller is that they are shown to be there by arguments involving unitarity. These phase relations are such that the quantity R, if one plots it as a function of A, is not a constant - as was sort of envisaged. But if in Γ/D as a function of A one has certain giant resonances, the function R goes sort of like the "principal value" to that. And the final part of my story is that within the Kapur-Peierls formalism the strength function $\pi\bar{\Gamma}/D$ has the $1/v$ behaviour of the old resonance scattering which everybody was brought up on, and R is the term of next order in k. Because of the complex boundary conditions in the Kapur-Peierls theory, one has to take into account the imaginary parts of these functions, which are complex, in order to get this, and I would point out that this is a problem that I'm sure is soluble within the Kapur-Peierls theory but nobody has worked it out. It's good that there's something left to do!

References

(1) P. L. Kapur and R. E. Peierls, Proc. Roy. Soc. A166 (1938) 277.

(2) Methods and Problems of Theoretical Physics, in honour of R. E. Peierls, Ed. John Bowcock, North-Holland 1970.

(3) T. F. O'Dwyer, M. Kawai and G. E. Brown, Phys. Lett. 41B (1972) 259.

6

Phase Transitions

D. J. THOULESS

Continuing this session of people who've spent ten years in Birmingham, I want to talk this morning about some developments in the theory of phase transitions for which Peierls was responsible, or with which he was associated, and to show how these ideas have developed and have come very much into their own in recent years. I shall be talking particularly about the effect on statistical properties of dimensionality, the number of spatial dimensions of the system one is considering. There was a quite extensive exploration of this forty years or so ago, and I think at that time it was probably largely irrelevant to the experimental situation. These days this is not the case. There are a number of physical systems which are essentially systems with low dimensionality in one sense or another. The sense is not always quite the same, but I don't have time to go into detail on this.

I will show you first of all (Fig. 6.1) - I could make a much longer list - some of the systems which have been studied in recent years. I would draw your attention to the first two initially; helium films on graphite or vycor and films of neon and argon. There are particularly nice experiments on helium films and neon films on graphite which have been done at Seattle, and I know from talking to people involved in this how much Peierls has encouraged them, and discussed with them this particular work. So this is something that he is associated with at present. I mentioned superconducting films, and layer compounds. These are two very different sorts of systems. In the one case, you have just single film on top of substrate. In the other case, you have layers of metallic atoms separated by non-metallic layers - you can squeeze organic compound in between the layers to get a very large separation so that conduction is essentially just in the layers and not

Dimensionality 2	He films on graphite or vycor
	Ne films
	Superconducting films - Bi, etc.
	Superconducting layer compounds $NbSe_2$, etc.
	Magnetic layer compounds - Rb_2MnF_4, etc.
	Inversion layer in semiconductor
Dimensionality 1	He in narrow channels
	Superconducting whiskers
	Type II superconductor in magnetic field
	Magnetic chains - $N(CH_3)_4MnCl_3$
	$K_2Pt(CN)_4Br_{0.3}$
	TTF - TCNQ
Dimensionality 0	Superconducting granules

Fig. 6.1. Some systems of low dimensionality.

between the layers. There are similar magnetic systems. One system which I recently have been very much interested in, prompted by Mott, is the question of the inversion layer in the semiconductor where you have a semiconductor-insulator surface. You can pump in electrons, which get trapped in that surface by their own image charge and you get a system 10 or 15 Å thick, containing a charged layer which can be seen apparently to undergo fairly dramatic transitions between insulating and conducting states. This in fact is regarded by Mott as being a particularly nice piece of evidence for this abrupt change in the conductivity when you go from a conducting to an insulating state, which Professor Edwards mentioned yesterday, but to some extent dismissed.

Below that in the table (Fig. 6.1) I've listed some one-dimensional things. These are systems which are so narrow that the only variation of whatever characterises the phase transition that one is concerned with, the only variation of significance, is the one-dimensional one along the narrow channel or along the superconducting whisker, and another case which I have got particularly interested in recently is the type II superconductor in a magnetic field. The magnetic field essentially constrains the electrons to move along the direction of field lines and the system behaves in many respects like a one-dimensional system, although, of course, basically it is a bulk system. Of course, if one can get a one-dimensional system, one has somehow to get enough of it that one can observe it and that always means that it is liable to be in some respects bulk - that is, in the case of the helium for example, you can't just have a single very narrow channel 20 or 30 Å across, and observe the flow of helium in that, because there is far too little of it. You have a large number of parallel channels and observe this. The ones that I have put at the bottom of this list are particularly dramatic, the platinum complexes which I shall mention a bit later. These are essentially chains of platinum atoms which behave metallically along the chain and are good insulators perpendicular to the chain. So much so, I am told, that, if you observe it in polarised light, it looks metallic in one direction of polarisation and looks more or less transparent in the perpendicular direction of polarisation. Then there are these organic compounds such as the TCNQ-based compounds (which is tetracyanodiquinomethane, I think, if I've got it right).

These are all one-dimensional systems and one also talks - I won't go into this at all - about essentially zero-dimensional systems which are so small that there is not room for variation of the superconducting wavefunction in the granule, that is, it is constrained by the boundary conditions but this also means, of course, it is so small that fluctuations are very important. So it's not something just described by mean field theory; it's something described by fluctuation theory. I won't go into detail on any of these; a number of them will come up later in my talk. To show where the subject has developed from, let me start with a quotation from one of the most quoted papers in statistical mechanics, namely, Ising's paper on the Ising model in which he correctly shows that the one-dimensional Ising model has no ferromagnetic properties:

> "Es wird gezeigt, dass ein solches Modell noch keine ferromagnetischen Eigenschaften besitzt, und diese Aussage auch auf das dreidimensionale Modell ausgedehnt."
>
> E. Ising, 1925.

He says that this can be extended to the three-dimensional case and there is some sort of argument which purports to do this, and it is a most important question, of course. Statistical mechanicians have spent a very long time, since this paper was published, in discussing the ferromagnetic properties of the three-dimensional systems without ever coming to a nice clean answer. So that's the starting point, and I want to show how some of the developments of the 30's showed that dimensionality can't be swept away by this sort of simple statement.

Since I've calculated it recently, because of the interest in the superconductor, let me just show you what the specific heat looks like for a one-dimensional system. (Fig. 6.2).

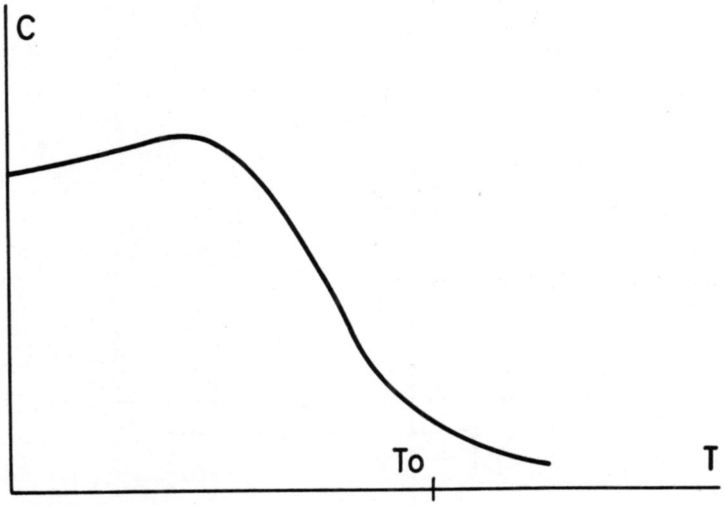

Fig. 6.2. Specific heat of a one-dimensional system.

The origin of temperature is arbitrary - I mean, one supposes it's a long way off to the left of this diagram. What you get is a specific heat which, looked at on a large scale, would have a discontinuity. You go from a high constant value in Fig. 6.2 to a lower value but it doesn't go abruptly. It rises slightly and then drops off smoothly over a temperature range which is going to be determined by the parameters of the problem; T_o is the mean field theory transition temperature in this model.

So this is typical of one-dimensional phase transitions: there is no abrupt change in the specific heat; no rising to a singular behaviour that you get in two- and three-dimensional systems, according to calculations and experiments that have been done.

Now, in fact one of the first demonstrations - as far as I am aware, the first completely convincing demonstration that the two-dimensional Ising model was really different - was one published by Peierls in 1936, where he treated the two-dimensional system by looking at the boundaries - the boundaries between the spin-up and spin-down regions (Fig. 6.3). There are two possibilities shown in Fig. 6.3(a) and (b). In Fig. 6.3(a) the boundaries are

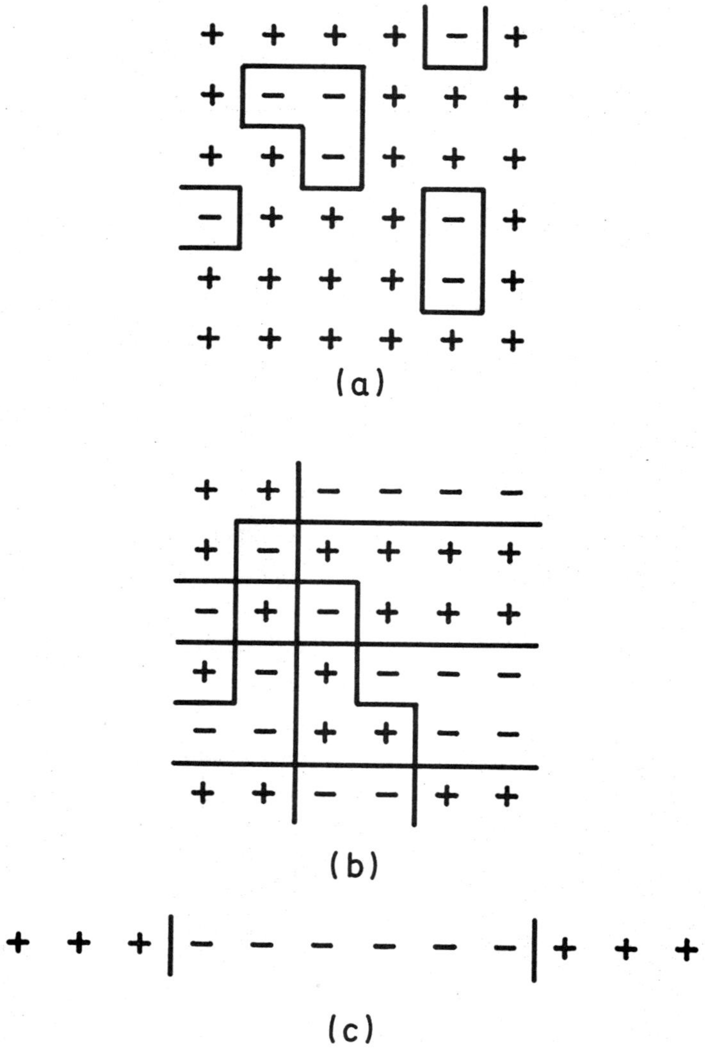

Fig. 6.3. The Ising model in two dimensions ((a) and (b)) and in one dimension ((c)).

drawn so that you have most of the spins up but a few clusters of spins are down. You can calculate or attempt to calculate both the energy, and the entropy of these boundaries and find out whether they are closed, as I've drawn them in Fig. 6.3(a), or whether they extend right through the system, as I have drawn them in Fig. 6.3(b). One can therefore go from the problem of spin-up and spin-down to an equivalent statistical problem of boundaries, and the method used by Peierls in this treatment was to consider the statistical mechanics of the boundaries, relax certain of the constraints on the boundaries - I mean, you mustn't have two boundaries coinciding - relaxing those constraints, and showing that still at low temperatures you've got, despite the relaxation of these constraints, still an ordered state, that is, the system in which there are no boundaries going right through the system. And, although it would seem that this proof had been made obsolete by the work of Onsager in finding the exact solution of the two-dimensional model, which is really a great landmark in the theory of phase transitions, in fact this is quite a lively subject, even at the moment. Much of the rigorous work on statistical mechanics depends on just this kind of argument. In fact, I've seen published in recent years generalisations of this argument to systems which are not covered by Onsager's proof, such as systems in which you have, as well as the nearest neighbour interaction, a next nearest neighbour interaction between the spins, which cannot be dealt with by Onsager's methods. In Fig. 6.3(c), I've drawn the one-dimensional system just to point out the obvious difference with the two-dimensional systems that, once you introduce a boundary, it necessarily goes right across the system and that the spins to the right of this boundary have no knowledge, are not influenced at all, by the fact that the spins on the left are spin-up, whereas, of course, in the two-dimensional system the region of spin-up is multiply connected, and so the introduction of another boundary does not completely destroy the order as it does in the one-dimensional case.

Slightly before this particular proof came out, there was the development of Bethe's method for resolving the Ising model, which was developed further by Peierls. It gives the one-dimensional results correctly, and gives a sort of spike in the specific heat in three dimensions, but not the kind of power law singular behaviour that is now regarded as so important and so characteristic of phase transitions. For that reason I think that the Bethe-Peierls method has become somewhat unfashionable. In fact, I looked at a recent fairly comprehensive book on the subject of statistical mechanics and phase transitions published by a University Press - which I won't mention in this room - which makes no reference either to Bethe or to Peierls in the fairly extensive bibliography. I think, in fact, the ideas of the Bethe-Peierls method, which are that you treat a finite cluster exactly, taking into account the boundary conditions, saying that this cluster is embedded in an average field provided by the atoms of the boundary, this method does suffer from the drawback of mean field theory, that is, that when you get close to the critical point everything gets, in fact, determined by what you assume about the boundary. Therefore, it does not differ all that much, or it suffers from the same drawbacks, as the mean field theory. Only these drawbacks matter in a very much narrower temperature range. However, you can't get a power law out of this sort of treatment. I think a logical extension of this method is not to treat the boundary conditions as giving you a fixed field, but to have a statistical variation, and to determine that statistical variation in some way self-consistently. This is really the basis of the renormalisation group ideas which are so fashionable at the moment.

Renormalisation group also brings in this idea of dimensionality, but there the easy problem is the four-dimensional problem. The reason for this (people concerned with field theory would know about this) is connected with the fact that the infrared divergences are only logarithmic in four dimensions whereas they are much more serious in a lower number of dimensions, and it is the infrared divergences, in fact, that one is concerned with in the problem of phase transitions. I'll just quote one result from this approach. As well as dealing with things like the Ising model and the Heisenberg model, you can also deal with the problem of the non-intersecting random walk, where the mean square distance gone in a walk of length ℓ is proportional to some power of ℓ, for large enough ℓ, and everybody agrees that the power is one in four dimensions, and it is two in one dimension, because in one dimension the only non-intersecting walk you can go on is a straight line, and therefore $r^2 = \ell^2$. Florey's theory and Edwards' work and other similar lines suggest that the exponent is 1.2 in three dimensions and 1.5 in two dimensions, or if it is not that, it is very close to it. This gives you the formula

$$r^2 = \ell^{6/(d + 2)} \tag{6.1}$$

where d is the dimensionality. Now, when the renormalisation groups are applied, you come out with the exponent 2ν, which is the same as the exponent of ℓ in the above equation

$$2\nu = 1 + \varepsilon/_8 + 15\varepsilon^2/_{2^8} \qquad (6.2)$$

where ε is the difference between four and the dimensionality of the system, and therefore in three-dimension ε is 1. These are the first terms of a power series.

Now, although if you substitute ε = 1 you get 1.18, and so very close to 1.2, and it's easy to believe that it's trying to tell you that the answer is 1.2, this renormalisation group result is in conflict with Eq. 6.1, which can also be written as

$$2\nu = 1/_{(1 - \varepsilon/_6)} \qquad (6.3)$$

So if the renormalisation group people are right, one should not be able to get the formula (Eq. 6.1) as an exact result. So I don't know what the status of that is going to be, but that's as far as I know, where things stand at the moment.

I want to talk at least briefly about one or two more things, one that I have been very much concerned with, again an idea that Peierls developed in the first place and published in 1934, concerning systems in which the spin is not just up or down, but can have a direction, so it can go continuously from up to down over a distance. In the Ising model it just has to go between up and down in one atomic layer, but of course if you have a vector that can turn round over quite a large distance, this has the effect of reducing the surface energy, and one can show that very similar considerations apply to the question of solid state order - the difference between the solid and the liquid - because you can move from one lattice position to form a crystal so that the lattice spacing is altered in a continuous manner.

Now, I'll give the argument for the case of the solid. What is shown is that in two dimensions or, even more strongly, in one dimension one cannot have solid state order in a rather strictly defined sense, namely, that you have a sharp Bragg peak. The argument goes in this way, I'll just run through it very briefly. If you look at the energy associated with a phonon mode of wavenumber $\underset{\sim}{k}$, you get the expression

$$\tfrac{1}{2}Nmu^2k^2A_{\underset{\sim}{k}}^{\;2}$$

where u is the velocity of sound and equipartition tells you this must equal $\tfrac{1}{2}KT$. Then you say, what's the mean square displacement of an atom from its equilibrium position, and that you can get in terms of the amplitude, and you end up with

$$<(x - x_o)^2> = \sum_{\underset{\sim}{k}} \frac{1}{2}A_k^2 = \frac{KT}{Nmu^2} \sum_{\underset{\sim}{k}} \frac{1}{k^2} \tag{6.4}$$

Now, if you are in three dimensions, that sum over $\underset{\sim}{k}$ is convergent when you replace the sum by an integral, you get something independent of the size of the system. The integral form is

$$<(x - x_o)^2> = \frac{KT}{(2\pi)^d mu^2} \int \frac{d\underset{\sim}{k}}{k^2} \tag{6.5}$$

In two dimensions it's logarithmically divergent with the size of the system, in one dimension linearly divergent. This argument can be applied in various forms, to solids, isotropic ferromagnets and antiferromagnets, and to superfluids.

Now, both experiment and calculation suggest that in two dimensions there probably are solids and superfluids - I mentioned these experiments in Seattle. There are other similar experiments on the helium films. Both this and certain calculations suggest that there really are solids and superfluids and this is a problem that both Peierls and I have been concerned about recently.

One way you could perhaps get round this argument in two dimensions is to say that, of course, in statistical mechanics we all like to think that we go to the limit of N infinite, but N infinite is not quite the same thing physically as log N infinite, and to say there is no solid is to say that this log term dominates. However, even if it is not infinite, it's quite an uncomfortably large number, and I am not happy about using this to get round these arguments. I should say, by the way, that the original Peierls argument was made rigorous by a group of people - model-independent I should say, which it certainly wasn't in its original form - on the basis of the Bogoliubov equality, by Mermin, Wagner, Hohenberg and a number of other people for different systems. So there is nothing in the theory that one can get away from.

The problem that I have been interested in - this was work with Mike Kosterlitz - was the possibility that, although the solid may not have order in the sense of a sharp Bragg peak, yet maybe it has order in the sense that there is still a crystal lattice without dislocations. We've built very much on the dislocation theory of melting in this work. To show what I mean, I've drawn in Fig. 6.4(a) a lattice in which I've made a continuous displacement of the atoms, so that those atoms at the top right corner of the picture are displaced by a whole unit cell relative to those at the bottom left corner.

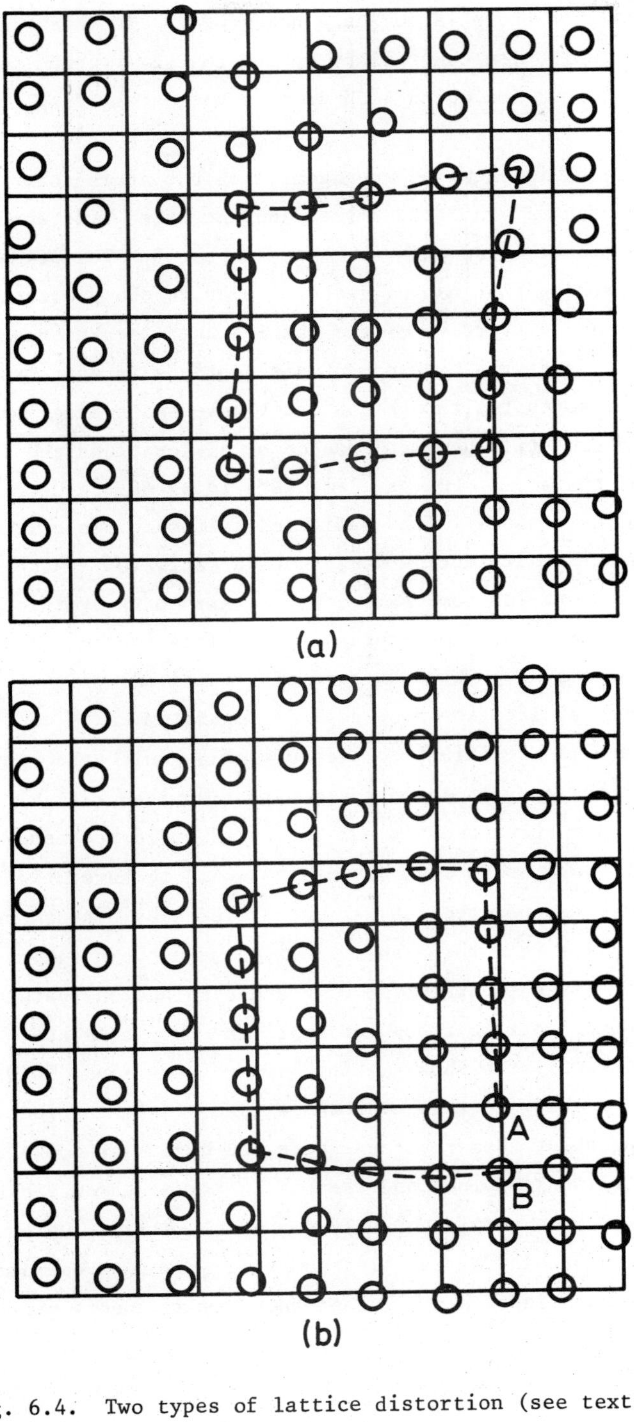

Fig. 6.4. Two types of lattice distortion (see text).

This is the kind of displacement envisaged in Peierls' argument and is sufficient to ensure that there is no sharp Bragg peak. In this picture there is no dislocation, which you can tell by trying to draw rectangles. I have shown a dotted square for which each side is four atoms long, and which is closed.

In Fig. 6.4(b) is shown an arrangement of the atoms that contains a dislocation. The "square" fails to close, and the first side starts at the atom A, but the fourth side ends at its neighbour B. The presence of a dislocation is something that can be detected even in the presence of the kind of displacement and distortion of the lattice that the Peierls argument said must happen. In fact, the presence of dislocations is what is necessary to have fluid flow, so that although you can't have sharp Bragg peaks in two dimensions, you can still have an absence of dislocations and therefore no viscous flow of the solid. One's elementary ideas of the difference between liquids and solids would still be satisfied, although one's more sophisticated ideas built on X-ray diffraction would not be satisfied. Mike Kosterlitz and I have argued that at low temperatures there are no free dislocations in thermal equilibrium, since the energy of a dislocation in two dimensions increases logarithmically with the area of the system. Since, however, the entropy of a free dislocation also depends logarithmically on the area of the system, there is a critical temperature below which the energy dominates and above which the entropy dominates, so that at the critical temperature dislocations appear spontaneously and the rigid solid becomes a viscous fluid. If this is right, one would also expect the solid to have orientational order, which, as Peierls has pointed out, is not excluded by his argument. Orientational order implies that the displacement between two neighbouring atoms in Fig. 6.4 is correlated in direction with the displacement between another pair of neighbours, no matter how far apart the two pairs are.

I'll just briefly finish by making some remarks about TCNQ and the platinum compounds - I think at this conference I shouldn't conclude without saying something about them - because here you have what looks like one-dimensional metallic behaviour and there's an argument you can find in Peierls' book which is very frequently quoted these days, which says that it's energetically favourable for the electrons and ions in a one-dimensional system to be correlated in such a way that there is an energy gap at the Fermi surface. I'll just briefly sketch what I regard as the reason for that. Fig. 6.5(a) is the picture for two-dimensions. The Fermi surface for an

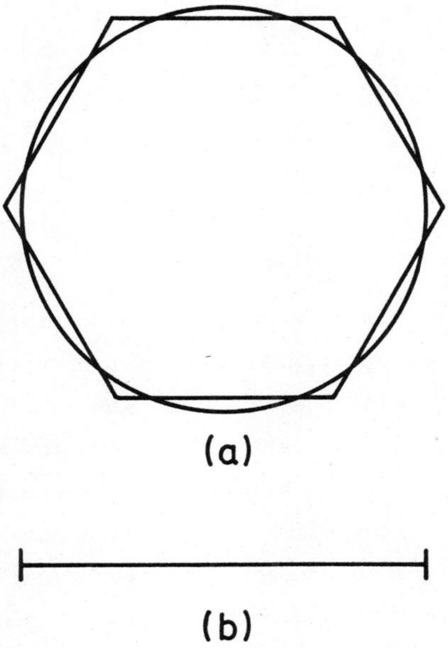

Fig. 6.5. The Fermi surface for an insulator. (a) two dimensions, (b) one dimension.

insulator is a Brillouin zone, and one possibility, of course, is a hexagon which is a reasonably close approximation to the circle you get with a free electron model. But still you have to pay in kinetic energy to distort the Fermi surface into the shape of the Brillouin zone, and that energy you have to pay if the thing turns itself into an insulator. In one dimension you don't get that happening, because the Fermi surface is just two points on a line (Fig. 6.5), and you can't distort that, so it costs less energy to make an insulator in one dimension than in two or three dimensions. There has been a great deal of excitement about this over the past couple of years. With TTF-TCNQ you get a rise in conductivity as temperature is lowered, which suggests that a superconducting state might occur at about $59^{o}K$. However, at a degree or so higher than that in temperature, the thing suddenly becomes insulating - bad luck. This is interpreted as the Peierls transition, or possibly the Peierls/Fröhlich transition, since Fröhlich's theory of superconductivity in 1954, I think it was, had a very similar mechanism in it. Bardeen has argued recently that in an ideal crystal with no defects, the Fröhlich solution would indeed be correct, and the Peierls insulator would, in fact, be a semiconductor, that is, the whole charge can move along carrying its ions along with it. Just to emphasise how fashionable this particular line is at the moment, let me say that there arrived in may mail yesterday yet another paper called "Some fluctuation and electrodynamic properties of the Peierls-Fröhlich conductor".

Discussion

Peierls: It is always very pleasing when things one was interested in long ago turn out to be still relevant, and sometimes even useful, and that comment goes for what Gerry Brown said as well as for what David Thouless has told us. However, I think I should like to make one reservation about the last topic in David Thouless' talk. I do not believe that the considerations about the behaviour of a hypothetical one-dimensional system have much relevance to what is found in the kind of substances which have long strings of atoms rather far apart, and which therefore in one sense resemble one-dimensional systems. As you pointed out, the difference between the one-dimensional and the two- or three-dimensional systems is that in the former case you can get an exact coincidence of the Fermi surface with the edge of the Brillouin zone. Now, if you have a very anisotropic but still three-dimensional system, your Fermi surface is still something like a sausage, something like an ellipsoid, and hence very distorted, and your Brillouin zone is still straight. So, you still cannot get the two to coincide. Now, it can of course happen that they can be close enough together, so that it pays to make the transition, but that can also happen in three dimensions; in fact, it does happen in three dimensions, as was shown very much longer ago in a paper by H. Jones. This work is somehow very unfashionable, but I think it gave the right explanation for the structure of Bismuth, and also, I think it can explain these structures much more than can a comparison with the strict one-dimensional model. I do not know whether you would agree with that? I think I detected some indication that you felt that way too - that these long string crystals are not really one-dimensional, for this purpose.

Thouless: I don't think I'd really like to comment on that. I have some feeling that agrees with that, but I have not really looked into these things closely enough to have a useful opinion. There is undoubtedly the fact that, in considering any sort of metal-insulator transition, one also has to take into account the correlation energy, and a one-electron model or even such a model supplemented by a Hartree-Fock type of argument is not going to be adequate.

Peierls: And also the zero-point motion of the lattice may be quite important as well. May I finally comment also on the question of whether there is an ordering possible in two dimensions. I entirely agree with everything you said about the theory, which certainly changes the image one had on the basis of the simple early arguments. I think we cannot yet accept any experimental test of that, both because of the logarithmic behaviour which is important in finite systems - these logarithms are very finite in any practical case - and also because I think there exists no two-dimensional experiment in which your atoms are not on some sort of substrate, which itself has a structure which may well help the ordering; in many cases the pattern that is formed seems to mesh in with the substrate. One cannot ignore that. So, I entirely agree, that is what the theory says and what any real test will no doubt show, but I do not think we have seen that test yet.

Elliott: Presumably we could paraphrase this by stating that Peierls is not really unstable? - In the sense that the prediction of the distortion caused by fitting these things on to the Fermi surface would cause a phase transition in one dimension, which you already showed would not exist. At finite temperatures the entropy always beats you in a one-dimensional system. What you really have here is a high tendency towards this kind of phase transition, which is actually driven in the end by the three-dimensional aspect of the system. I really wondered if you could comment on how sensitive it was to the three-dimensional nature. I mean, does it somehow go exponentially with the interaction into the other two dimensions?

Thouless: I certainly do not know the answer to that in the case of the one-dimensional transition. My impression was that probably in an ideal system it would behave in some respects very much like a one-dimensional system, but always of course there comes a point where one gets so close to a transition that correlations between chains are going to come into play. Once you start getting correlations coming in as fluctuations, these can very much enhance the initially very weak interaction between neighbouring strands. This is certainly true in these magnetic chain problems where, when you get close to the transition, there is a rather long range of magnetic order within a chain;

this means that the interaction between chains is coherent and the system will suddenly behave in a three-dimensional way, so that you do get a sharp phase transition. I suspect that there is an analogous situation here, although I do not know it for sure.

Edwards: Can I make a comment about the question of extrapolating in dimensions? As you said correctly, you get in a low number of dimensions (1, 2, 3) infra-red divergences in correlation functions, which appear in statistical mechanics. That means that if you work in 5, 4 ... dimensions downwards, you can produce solutions, which you can attempt to extrapolate down into three dimensions, so as to yield the series of the kind you wrote down. Now, if you start the other end and work up in dimensions, you find ultra-violet divergences; that is, for example, in the non-intersecting random walk, you find that the effective field set up by the walk as it progresses goes like $1/R^3$ in five dimensions. So there is an ultra-violet divergence there. So you can attempt to start in a high dimension and work down, or, by using self-consistent fields, you can start in a low dimension and work up. Now, if you try these extrapolations, both of them give violently divergent series, and the fact that the two series don't seem to have any resemblance to one another is not surprising, since they both diverge. And what the truth is, I do not know. But three dimensions sits somewhere between the one kind of divergence and the other kind of divergence, and, in fact, is entirely handleable and entirely physical, as we live in that number of dimensions!

Thouless: I think I have one comment on what Sam just said. I don't think that the renormalisation group methods really involve extrapolating from above four dimensions. In four dimensions there are logarithmic terms. These are completely ignored by this kind of approach, and I think it would be claimed that it is for dimensionality just less than four that things are simple. If you take dimensionality much greater than four, you get into a completely different situation which has no relevance at all to the three-dimensional case. What is hoped is that on this low side of four dimensions, where you have got infra-red divergences, but the ultra-violet divergences can be ignored, you have a situation which has some relevance to the physical situation. At the moment, I think, the arguments in favour of this are mostly numerological ones; that is, the numbers which have been calculated are on the whole acceptable numbers. My impression is that recently people have come out with a little more - that is, that the method has done more than reproduce badly things that were known, for example, from Domb's work, but has actually taken the theory a bit further. This case I gave, I think, is one case in point, and it would be worth having a very careful look at that to see whether it is actually right; there seems to me a definite conflict between what one might hope from looking at one, two, three and four dimensions, and what the renormalisation group method gives as a firm prediction.

7

Heavy Particles

R.H. DALITZ F.R.S.

During the last year or so, many proposals have been put forward, which require the existence of various kinds of heavy particle. Some theoreticians have come to feel very strongly about the reality of their particular proposal and its heavy particle. The question is whether such particles will provide new fields for fruitful research in the future, or whether their proliferation is the symptom of a disease. We shall not be able to decide this question this morning. My aim is simply to sketch the reasons why some of these proposals have come to seem so attractive, and to comment briefly on the chances for settling some of these questions in the not-so-distant future.

Let us start with the weak interactions. Consider beta decay, or the corresponding neutrino-induced reaction

$$\nu_e + n \rightarrow p + e^- . \qquad (1)$$

This is fairly well understood, and we represent it normally by the four-particle vertex given on Fig.7.1 below. This involves an interaction between two charge-changing currents, a lepton current and a hadron weak current. The interaction depicted in Fig.7.1 is point-like, with a coupling strength G whose value is about $10^{-5}/M_p^2$ and which is generally referred to as the Fermi coupling constant. From far back in the distant past, I can recall

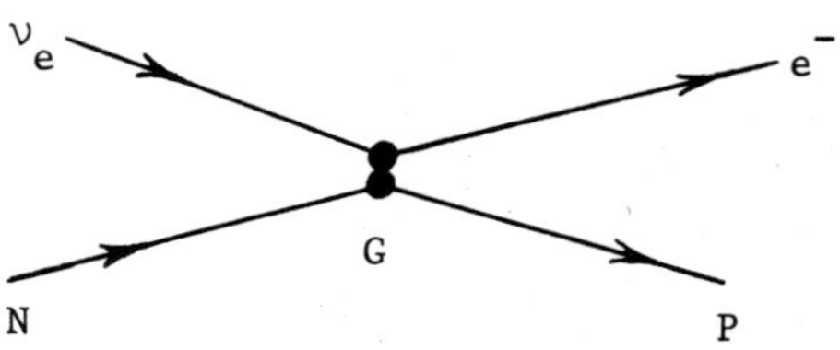

Fig.7.1. The beta decay interaction vertex

Professor Peierls remarking again and again that this theory could not be correct, for the following reason. If you take this interaction and calculate cross sections, you find that the cross section at high energies is proportional not only to G^2 but also to the square of the centre-of-mass energy E_c. This follows simply from dimensional considerations: G has dimensions $(\text{mass})^{-2}$ and occurs squared, the cross section must have dimension $(\text{mass})^{-2}$, (when we use natural units, $\hbar=c=1$), and the only relevant mass in the limit $E_c \to \infty$ is that corresponding to E_c itself. For the ν_e-N interaction, for example, we have $E_c = \sqrt{((E_L+M_N)/2)}$, where E_L denotes the incident neutrino energy in the laboratory frame (i.e. stationary N target) and the cross section is predicted to rise linearly with E_L. On the other hand, since it is a point interaction, it can be effective only in the s-wave, which implies that the cross section is bounded by $4\pi\lambda\!\!\!^{-}{}_c^2$, which goes like $1/(M_N E_L)$ in the high energy limit. Obviously, the calculated cross section is not consistent with this bound. Ultimately, for sufficiently high energy, the two formulae are incompatible, and so Professor Peierls' conclusion was that one should not believe the theory. In fact, in recent years, the experimental data on neutrino cross sections has become quite extensive. For incident neutrino energies up to 200 GeV, the neutrino-proton cross section does vary linearly with E_L, as the simple theory predicted, but of course its value still lies far below the geometrical limit $4\pi\lambda\!\!\!^{-}{}_c^2$.

The close relationship, first pointed out by Feynman and Gell-Mann, betwen this hadron weak current and the isospin current for hadrons, the latter being part of the electromagnetic current for hadrons, suggested that there may exist some intermediate field which carried the weak interaction, as the photon carries the electromagnetic interaction. This field is now referred to as that of the W boson; W comes in positively and negatively charged varieties, at least, and it is a vector particle, as is the photon. In this case, Fig.7.1 becomes replaced by Fig.7.2, showing the weak interaction mediated by the W boson field. The W-exchange introduces an additional factor $g^2/(M_W^2-q^2)$, where q denotes the energy-momentum transfer, M_W is the W boson mass and g denotes the coupling constant for the interaction between the weak currents and the W field. In low energy experiments, q^2 is

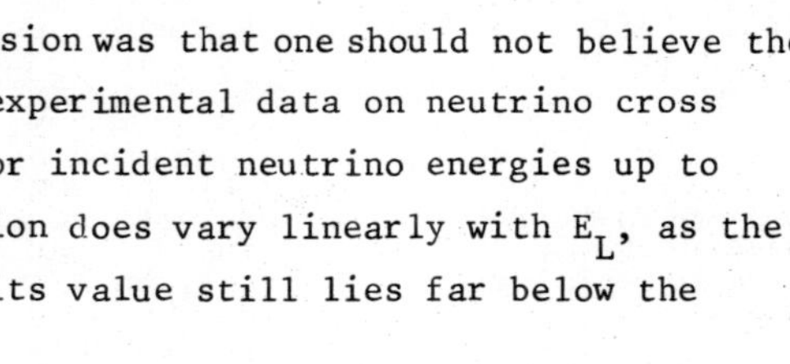

Fig.7.2. The beta decay interaction mediated by the exchange of a massive W boson

small and may be neglected in comparison with M_W^2, and so, comparing the formulae for Figs.7.1 and 7.2, one finds the relation $G \sim g^2/M_W^2$. In general, however, we must now take into account the finite range of the W-exchange interaction and the interaction falls in strength as the momentum transfer $(-q^2)$ increases. There was an early paper published by Professor Peierls and N. Byers discussing this W-exchange model.[1]

Some time ago, Salam and Ward proposed to take the analogy between the weak currents and the hadronic currents a step further[2]. They suggested that it would be logical to take the weak coupling constant g^2 equal to e^2 (the electromagnetic coupling constant $\approx 1/137$). This would lead to a conceptual unification of the W boson and the photon fields. However, this requires the identification of G with e^2/M_W^2, from which one must conclude that the mass M_W should be of the order of 37 GeV. At the time, this sounded a pretty odd suggestion. Firstly, it appeared peculiar to put the particles $W^{\pm}$ and γ into a multiplet (W^-, γ, W^+), when one of the particles was guaranteed to have zero mass, while the other two had mass of order 37 GeV. Secondly, at that time, we were much more impressed by the differences between the weak interactions and other interactions, rather than by any similarities. In particular, it was known that the weak interactions did not satisfy parity conservation or charge conjugation invariance, both of which were exceedingly well established for photons and for the electromagnetic interactions generally.

This picture has recently been revived by Weinberg in a somewhat different form, within the context of a gauge theory in which the W bosons are the gauge particles for the weak interaction currents [3]. We shall not describe this gauge theory in any detail here, but simply pick out the elements necessary. Besides the $W^{\pm}$ bosons, the theory begins with a corresponding W^0 boson and another neutral vector boson V^0. These two neutral bosons mix, resulting in two fields, one representing the photon and having zero mass, the latter feature being a consequence of exact conservation for the electromagnetic current, the other being the field for a massive Z^0 particle, whose mass is related with that for the $W^{\pm}$ bosons. The electromagnetic coupling now involves both the W^0 and V^0 fields, and so the relationship between the W coupling constant and the Fermi coupling constant G is not quite the same as that envisaged by Salam and Ward. This relationship depends on a Weinberg mixing angle θ_W, to be determined from experiment; explicitly, the relationship is

$$G/\sqrt{2} = (e^2/M_W^2)(\pi/2\sin^2\theta_W). \tag{2}$$

The charged boson mass now comes out somewhat larger than the estimate given by Salam and Ward, the result from (2) being

$$M_W = (37\ \text{GeV})/|(\sin\theta_W)|, \tag{3}$$

from Eq.(2) and the known values for G and e^2. Present empirical estimates[4] of the Weinberg angle now give $\sin^2\theta_W=0.35\pm0.04$, so that the Z^o boson should have a mass at least comparable with that for the $W^\pm$ bosons. Explicitly, the Z^o mass is given by

$$M_Z = (73\ \text{GeV}/|(\sin 2\theta_W)| \approx 77\ \text{GeV}. \tag{4}$$

This gauge theory, unifying the weak and electromagnetic interactions, implied the existence of weak neutral currents with which the Z^o field couples, just as the photon couples with the electromagnetic current. The existence of such weak currents has now become well established. (An important point, to which we shall return below, is that the neutral weak currents which have been observed involve no change of strangeness). Of course, one has to realize that the mere existence of neutral weak currents does not imply that this particular form of the theory is correct; such currents could arise in many other ways. However, it is at least an encouraging outcome. Further, we should remark that it was already well known that a theory with vector gauge particles of zero mass coupled with conserved hadronic currents is necessarily renormalizable. At the time of his proposal, Weinberg speculated that this renormalizability would remain the case through his transformation from a Lagrangian with zero mass gauge particles to a final Lagrangian describing massive gauge particles (a mechanism known technically as "spontaneous symmetry-breaking"), and this was later demonstrated to be so by t'Hooft.[5] With such a renormalizable theory, we are in a position at last to make finite and meaningful calculations for weak interaction processes and indeed for all the phenomena which are induced by the weak and electromagnetic interactions together. For this reason, this unified theory has been very attractive to theoreticians and it has spurred them into great activity and much enthusiasm. However, it has not really been tested quantitatively and one should not take it for granted that its present form is necessarily the physically correct form. It is a simple member of a large class of theories which may involve a larger number of weak bosons and some number of heavier leptons, not yet observed. However, this class of theories does require the existence of such heavy bosons, and we do expect that charged W particles (and probably neutral Z^o particles also) will become established in due course.

The direct observation of weak bosons is not likely to be possible for some considerable time, because of the high mass values expected for them.

The c.m. energies available at present, at the CERN Intersecting Storage Ring (≈60 GeV c.m. energy) and at the Fermilab accelerator (proton laboratory energy up to 400 GeV, hence c.m. energy up to 28 GeV for proton-proton interactions), are well below the energies necessary for production of these particles. It is a reasonable extrapolation to expect that a 400 GeV proton storage ring might be built within about 10 years - it would be cheap at the price for what could be achieved with it - in which case the limitations on W and Z production would be a question of cross section rather than of energy available. Among hadron-hadron processes, the most favourable reaction would be W pair production, as soon as the c.m. energy available exceeds $2M_W$, since this is possible then through virtual electromagnetic interactions, whereas single $W^{\pm}$ or Z^o production necessarily involves the weak interactions. Single $W^{\pm}$ or Z^o production would be expected to be quite prominent among neutrino-nucleon interaction processes at sufficiently high neutrino energy, for example through processes such as

$$\nu_\mu + p \rightarrow \mu^- + W^+ + p \tag{5}$$

However, the c.m. energy corresponding to the energy (≈200 GeV) of the most energetic neutrino beams available at present is not much more than 20 GeV, so there is a long way to go before weak boson production can be studied through neutrino-induced processes such as (5).

This all seems very reasonable, and we can feel a certain degree of confidence that the question of the existence of weak bosons and of their properties will be settled within the foreseeable future. The weak bosons will decay through interactions involving their coupling with the weak currents, both leptonic and hadronic, the former giving the modes $W^+ \rightarrow e^+\nu_e$ and $\mu^+\nu_\mu$ and $Z^o \rightarrow e^+e^-$, for example, and the latter giving many multiparticle hadronic modes, such as $W^{\pm} \rightarrow \pi^{\pm}n\pi^+n\pi^-r\pi^o$ and $Z^o \rightarrow n\pi^+n\pi^-r\pi^o$, for integers n and r, and others. The lifetimes are expected to be intermediate between those known for the weak decays of hadrons (typically 10^{-10} sec) and those known for the hadronic decays of hadrons (typically 10^{-23} sec), thus in the region of 10^{-16} or 10^{-17} sec. Such lifetimes are rather long in relation to nuclear time scales, but rather short in relation to the scale of space and time measurements we are accustomed to make in the laboratory.

The next particle I want to mention is the quark, about which I shall say relatively little. The situation concerning the quarks as constituents of the hadrons has become more and more favourable as time has passed. Sum rules involving measured lepton-nucleon inclusive cross sections have been especially valuable. The data on deep inelastic electron (e→e'), neutrino ($\nu_\mu \rightarrow \mu^-$) and antineutrino ($\bar{\nu}_\mu \to \mu^+$) scattering on nucleons has been found to have the property of "Bjorken scaling". This has the consequence that all of the available data may be summarized by three structure functions $F_i(x)$ for i=1,2,3, where x denotes the kinematical variable $(-q^2)/(2M_N q_0)$, q_0 being the energy transfer from the lepton system to the hadrons, as seen in the laboratory frame, and $(-q^2)$ being the square of the magnitude of the four-dimensional momentum transfer from the lepton system. This variable x runs from 0 to 1, and the structure functions reduce to F_2 and F_3 if we assume that the SU(3)-triplet sub-units (i.e. quarks) within the hadrons have spin $\frac{1}{2}$, a constraint which is quite consistent with this lepton-hadron scattering data. All that we need to know for the purpose of our present discussion is that these functions $F_2(x)$ and $F_3(x)$ can be determined empirically, from measurements for the electron-nucleus and neutrino-nucleus systems, the nucleus being regarded as a collection of independent neutrons and protons, equal in number. Two sum rules have particularly direct significance. The first of these tells us the number of quarks minus the number of antiquarks within a nucleon. It takes the form

$$N(q) - N(\bar{q}) = \int_0^1 \bar{F}_3^{\,\nu}\, dx = 3.2 \pm 0.35, \qquad (6)$$

where $\bar{F}_3^{\,\nu}$ denotes the ν-nucleon structure function averaged over neutron and proton, and the empirical value has been deduced from the empirical data obtained at CERN with the Gargamelle chamber.[6] This value is in accord with the quark model picture of baryons as tightly-bound three-quark systems. The second sum rule gives us the mean square charge of the non-strange quarks, as follows:

$$\left(\int_0^1 \bar{F}_2^{\,e}\, x dx\right) / \left(\int_0^1 \bar{F}_2^{\,\nu}\, x dx\right) = \tfrac{1}{2}(Q_u^2 + Q_d^2) = 0.29 \pm 0.03, \qquad (7)$$

where Q_u and Q_d are the charges of the u and d quarks, the $I_z = +\frac{1}{2}$ and $-\frac{1}{2}$ components of the non-strange quark doublet. The empirical value agrees with the charge values given for the Gell-Mann-Zweig quarks, namely Q=(2/3, -1/3, -1/3) for the quarks (u,d,s), which require 5/18 for this ratio. These results add considerable weight to the quark model picture of the structures of hadrons, giving some physical reality to the quarks which had previously been only hypothesized in order to give a quantitative interpretation for the complicated spectroscopy observed for hadronic states.

Extensive searches have been made for these Gell-Mann-Zweig quarks, both in accelerator experiments and in Nature, but no quarks have been found, although their fractional charge value would give an unmistakeable signal.

This might only mean that quarks are very heavy particles, with masses greater than 30 GeV. If their masses were lower than this, quark-antiquark pairs could be produced at the CERN Intersecting Storage Ring, since the proton-proton collisions studied there have c.m. energy up to 60 GeV; however, no quark production has been detected there, despite rather thorough searches. On the other hand, the scaling properties observed for lepton-hadron interactions suggest that the quarks within hadrons should have very little mass, perhaps even zero mass. If the quarks have a large mass, this would provide a natural scale parameter and it would be difficult to understand the success of the scale-free parameter x in describing the deep inelastic phenomena in lepton-hadron interactions. Well, if quarks have essentially zero mass, why has quark-antiquark pair production not been seen? The theorists have been busy, and so we have a number of theories to account for this, all following the pattern generically known as "quark confinement". Probably the most developed scheme of this kind is the "bag theory" which has been discussed by Weisskopf, Johnson, Jaffe and their colleagues at MIT(7) and about which I shall say nothing further. There seems to be a reality about quarks now. They may perhaps be heavy particles, but probably not; in any case, they might well never be seen in the free state.

One particularly odd aspect of the quarks is their statistics. For the three-quark states which represent the nucleons, we expect the internal orbital motion to be S-state, since the nucleons belong to the ground configuration for the three-quark system. This ground configuration is known to comprise the JP=1/2+ baryon octet and the JP=3/2+ baryon decuplet. Let us concentrate on the latter, where it is clear that the spins of the three quarks are parallel and the unitary spins of the three quarks are parallel; in these parallel configurations, there is complete permutation symmetry in the quark labels. The space wavefunction must also have complete permutation symmetry; with L=0+, any antisymmetry would imply the existence of nodes in this wavefunction and these are forbidden for the ground state (under appropriate and reasonable assumptions). Further, the lowest four SU(6) x O(3) baryonic supermultiplets are well described by the quark shell-model, and the ground state baryon octet and decuplet both receive interpretation as members of the $(1s)^3$ configuration, which clearly has even permutation symmetry. Thus, we conclude that the complete wavefunction for the ground state baryonic supermultiplet, including space, spin and unitary spin variables, must have even permutation symmetry. This would suggest that quarks obey Bose statistics, but we know this is forbidden for spin 1/2 particles, by very deep principles.

In any case, Bose statistics for quarks would imply Bose statistics for protons, which is certainly not the case. The simplest conclusion is that there must be at least one further degree of freedom for quarks whose wavefunction provides the antisymmetric factor needed in the proton (more generally, baryon) wavefunction. The most straightforward proposal is that of a "colour variable", as suggested by Gell-Mann,[8] according to which each quark can occur in three states, which are termed red, white and blue. With these new states, it is clearly possible to make the required antisymmetric factor. This is given at once by the determinant

$$\begin{vmatrix} q_r(1) & q_r(2) & q_r(3) \\ q_w(1) & q_w(2) & q_w(3) \\ q_b(1) & q_b(2) & q_b(3) \end{vmatrix} \tag{8}$$

where the suffices identify the colour component and i=1,2,3 are the quark labels. This factor is unique, when the colour space is three-dimensional, as is assumed here. Thus we now have quite a proliferation of quarks! Previously, we were unable to find any of three quarks - now there are nine quarks we cannot find. Further, where are all the additional states associated with the new quantum number of colour? The simplest answer which has been proposed is based on the idea that the Hamiltonian for quark interactions may have invariance with respect to SU(3)-transformations in the three-dimensional colour space. This invariance property is generally referred to as SU(3)'-symmetry, and it implies that hadronic states may be classified further, according to their SU(3)' representation. Even with this simplifying assumption, the introduction of this colour space necessarily brings with it two additional additive quantum numbers and a corresponding multiplicity of colour states. According to our present discussion, all of the low-lying baryon states are necessarily SU(3)'-singlet, since the colour wavefunction (8) may be rewritten in terms of vectors in colour space, as follows:

$$\underline{q}(1).\underline{q}(2) \times \underline{q}(3), \tag{9}$$

where the underline here denotes a vector in colour space, and is clearly a scalar quantity in colour space, i.e. an SU(3)'-singlet. It is then a natural hypothesis to propose that all the observed hadronic states are SU(3)'-singlet states, as was suggested originally by Han and Nambu. In this case, since the SU(3)' factor is unique, there is no immediate problem of the prediction of a greatly enhanced multiplicity of states in the present region of investigation; the many other SU(3)' multiplets which will exist are then limited to some high mass region, not yet explored. This last hypothesis is dynamical in character, and it was pointed out by Lipkin[9] that the dynamical properties required would follow directly if it were assumed that quark-quark forces were mediated by the exchange of a colour octet of neutral gluons. The most attractive hypothesis is that these gluons are vector particles, and further, that they are the gauge particles associated with colour invariance, a point of view advocated energetically by De Rujula, Georgi and Glashow.[10] It is even conjectured now that "quark confinement" may also be a consequence of the colour gauge field, which may allow only SU(3)'-singlet configurations. This is reminiscent of the bag theories, which are particularly clever in that they arrange matters so that quarks can only be found in bags, and that bags can contain only SU(3)'-singlet configurations of quarks and antiquarks; thus, all colour non-singlet states are abolished by the premises of the bag theories. Other theorists expect that coloured states, i.e. states which are not colour singlet states, should exist but that they are difficult to observe because of their high mass values, perhaps 3 or 4 GeV, or that they are already being produced but have not been identified correctly.[11] Thus, there may be a new category of heavy particle, those which are colour non-singlet, but this is not certain - there is a diversity of views about this.

What would happen to such heavy, coloured states, if they were formed? There are really two views about this. In the Han-Nambu theory, the red, white and blue triplets do not have the same charge values. Hence, the electromagnetic current distinguishes between them and is therefore not an SU(3)' scalar, so that it can induce transitions between coloured and non-coloured states. If such coloured states were produced hadronically or electromagnetically, they could decay rapidly by γ-emission to the SU(3)'-singlet states we know already and they could well be difficult to detect, in this case. Coloured states could be produced rather directly by electron-positron annihilation reactions, since the latter necessarily pass through an intermediate virtual photon, thus $e^+e^- \rightarrow$ "γ" $\rightarrow$ hadrons. Possibly some of the resonance states excited with the electron-positron storage rings (SPEAR at SLAC (Stanford University), and DORIS at DESY (Hamburg)) may be coloured states, as argued by Feldman and Matthews,[11] but there is no real evidence yet to demonstrate this. Indeed, we should emphasize at this point that the charge values for Han-Nambu quarks are not consistent with the sum rule result (7). The other view is that the red, white and blue triplets are indistinguishable, although not identical; in particular, the three triplets have the same charge values and the electromagnetic current is an SU(3)'-scalar. It is conceivable that the weak interactions might distinguish between the three triplets, but this is not at all necessary (and it has not even been suggested, to my knowledge). On this view, the coloured multiplets might be quite separate from the SU(3)'-singlets (and from each other), and they are therefore rather an embarrassment; if the colour gauge theory leads to their abolition, this would be a great relief.

Next, there are the gluons themselves, neutral entities within the hadrons. Their presence is indicated by the following sum rule, which gives the fraction of the proton momentum which is carried by the quark-partons,

$$3\left(6\int_0^1 \bar{F}_2^{\,e}\,dx - \int_0^1 \bar{F}_2^{\,\nu}\,dx\right)/4 = 0.46\pm0.2. \tag{10}$$

This implies that about 50% of the proton momentum must be due to neutral partons within the proton, partons which do not interact with the photon in any way, and it is natural to identify these with the gluons. If the gluons g are vector particles and have an SU(3)'-singlet component $g_V(\underline{1}_c)$, then it would be reasonable to expect gluon excitation by electron-positron annihilation

$$e^+e^- \to "\gamma" \to g_V(\underline{1}_c) \to \text{hadrons}, \tag{11}$$

followed by the allowed decays of $g_V(\underline{1}_c)$ to SU(3)'-singlet hadrons, through strong interactions, just as the excitation reactions $e^+e^- \to \rho^o$, ω^o and ϕ^o have been studied for the low-lying vector mesons. There have been suggestions(12) from the study of the proton form factor at high momentum transfer, that there might exist a heavy meson analogous to $g_V(\underline{1}_c)$ with a strong coupling to the proton and with mass of order 7-10 GeV. Similarly, Drell and Chanowitz(13) and West (14) have suggested that partons (quarks) may have an electromagnetic structure, and this could be attributed to the coupling of $g_V(\underline{1}_c)$ with the quarks. However, there is no direct evidence for this vector meson in the reaction (11) up to mass 7.5GeV, the top energy of SPEAR, apart from the possibility that the broad resonance ψ(4.4 GeV) might be a candidate. However, the width observed for ψ(4.4) is only about 200 MeV and nothing is known about its coupling to the nucleons. At the present time, ψ(4.4) is given quite a different interpretation (see below), and there is little theoretical motivation for such a gluon as $g_V(\underline{1}_c)$. Our theoretical prejudices would prefer the gluons to be members of a colour octet $g_V(\underline{8}_c)$, and the photon to be a colour singlet, in which case the gluon states cannot give rise to any resonance in the reaction (11).

Another class of particles, concerning which many theoreticians feel very strongly, has arisen in the context of the quantum number "charm", first introduced by Bjorken and Glashow in 1964 and which has been taken up again by Glashow, Iliopoulos and Maiani[(14)] in connection with gauge theories of the weak currents. These currents may be characterized by symbols such as $(\bar{u}d)$ and $(\bar{u}s)$, by which we mean that the current induces the $\Delta Q=-1$, $\Delta s=0$ transition $u\to d$, and the $\Delta Q=-1$, $\Delta s=-1$ transition $u\to s$, respectively; the hermitian conjugate currents must also be present in the Hamiltonian, $(\bar{d}u)$ inducing the $\Delta Q=+1$, $\Delta s=0$ transition $d\to u$ and $(\bar{s}u)$ inducing the $\Delta Q=+1$, $\Delta s=+1$ transition $s\to u$. From the empirical study of semi-leptonic decay processes, which are induced by the coupling of these currents with the leptonic currents, we know that the weak hadronic currents for the charge change $\Delta Q=-1$ have the Cabibbo form

$$\{(\bar{u}d)\cos\theta_C + (\bar{u}s)\sin\theta_C\}, \tag{12}$$

where θ_C is known as the Cabibbo angle, its numerical value $\sin\theta_C=0.23$ being well established now. This form (12) leads naturally to the idea that the weak interactions pick out from the quarks the particular doublet

$$\{u,\ (d\cos\theta_C + s\sin\theta_C)\}. \tag{13}$$

With this view, it is a natural supposition that the neutral weak hadronic currents should consist of the terms $(\bar{u}u)$ and

$$((\bar{d}\cos\theta_C+\bar{s}\sin\theta_C)(d\cos\theta_C+s\sin\theta_C))=\cos^2\theta_C(\bar{d}d)+\sin^2\theta_C(\bar{s}s)+ \\ +\sin\theta_C\cos\theta_C((\bar{d}s)+(\bar{s}d)). \tag{14}$$

As we said earlier, the effects of neutral weak currents have been observed in many neutrino experiments, and the expectation of their occurrence was part of the Weinberg-Salam-Ward gauge theory. However, the neutral currents whose effects have been observed are only those involving no change in strangeness. The neutral weak interaction processes involving $\Delta s\neq 0$ are known to have exceedingly small rates. Let us consider several cases. The $\Delta Q=0$, $\Delta s=-1$ decay process $K^+ \to \pi^+e^+e^-$ has branching ratio 3×10^{-7}, which is to be compared with the branching ratio 5×10^{-2} for the rather similar $\Delta Q=-1$, $\Delta s=-1$ decay process $K^+ \to \pi^o e^+\nu_e$. The former process is weaker than the latter by more than five orders of magnitude, and there is actually no real evidence that the $\Delta Q=0$ decay observed is due to neutral weak currents (the observed rate could well be due to the non-leptonic decay interaction $K^+ \to \pi^+ + "\gamma"$, with internal conversion $"\gamma" \to e^+e^-$ as a subsequent electromagnetic effect). Another example is given by the $\Delta Q=0$, $\Delta s=\pm1$ process $K_L^o \to \mu\bar{\mu}$, for which the partial decay rate is known to be 0.2 sec^{-1}. This is to be contrasted with the $\Delta Q=-1$, $\Delta s=-1$ decay process $K^+ \to \mu^+\nu_\mu$, which has partial decay rate 5×10^7 sec^{-1}. These comparisons provide evidence that the $\Delta s=-1$, $\Delta Q=0$ current is very much weaker than the

$\Delta s=-1$, $\Delta Q=-1$ current in (12), whereas the $\Delta s=\pm1$, $\Delta Q=0$ currents appear in expression (14) with coefficient $(\cos\theta_C \sin\theta_C)$, comparable with that $(\sin\theta_C)$ in expression (12). Hence, the theory must be modified.

The most economical modification is that due to GIM.[15] These authors invoke a fourth quark, which we may denote by c, which is an isospin singlet and carries one unit of a new quantum number C, called "charm". Thus C=+1 for the new quark c, and C=0 for the old quarks (u,d,s). The Gell-Mann-Nishijima relation must be amended to read

$$Q = I_3 + \tfrac{1}{2}Y + \tfrac{1}{2}C. \tag{15}$$

GIM then proposed that the weak interaction may also pick out a second doublet,

$$\{c,\ (-d\sin\theta_C + s\cos\theta_C)\}. \tag{16}$$

This doublet forms the $\Delta Q=-1$ weak current,

$$\{-(\bar{c}d)\sin\theta_C + (\bar{c}s)\cos\theta_C\}, \tag{17}$$

and contributes to the neutral weak current the terms $(\bar{c}c)$ and

$$((-\bar{d}\sin\theta_C + \bar{s}\cos\theta_C)(-d\sin\theta_C + s\cos\theta_C)) = \sin^2\theta_C(\bar{d}d) + \cos^2\theta_C(\bar{s}s) - \\ -\sin\theta_C\cos\theta_C((\bar{d}s)+(\bar{s}d)). \tag{18}$$

When these two contributions (14) and (18) are added together, the two terms $((\bar{d}s)+(\bar{s}d))$ cancel, and the empirically absent transitions $\Delta Q=0$, $\Delta s=\pm1$ are then absent from the theory. The net $\Delta Q=0$ weak current then has only terms with $\Delta s=0$. This is an appealing construction, and gives a serious motivation for considering the possible addition of a fourth quark c.

The subsitution of this quark c (or antiquark $\bar{c}$) for an ordinary quark (u,d,s) (or ordinary antiquark $(\bar{u},\bar{d},\bar{s})$) in a hadron will lead to a charmed hadron, with C=+1 (or C=−1). Since ordinary hadrons have C=0, the production of charmed hadrons in hadron-hadron collisions involves the creation of an intermediate $(\bar{c}+c)$ pair, and therefore final production of two hadrons with opposite charm values, C=+1 and C=−1, respectively, i.e. meson-meson or baryon-antibaryon pair production or meson-baryon associated production. Charmed particles could also be produced by neutrino-nucleon collisions, since the weak current formed from the doublet (16) includes the necessary transition $d\to c$. This last cross section might be somewhat low relative to those for other neutrino reactions (based on the transition $d\to u$ in the weak current (17) for example), since the $\Delta C=+1$ transition is weaker by the factor $\sin^2\theta_C \simeq 0.05$, but reactions such as

$$\nu_\mu + p \to \mu^- + B^{++}(s=0,\ C=+1) \tag{19}$$

will at least be a non-negligible fraction of the total neutrino-proton cross section. What mass should these charmed hadrons have? Some of the cancellations mentioned above (e.g. the low rate observed for $K_L^o \to \mu\bar{\mu}$) are exact only when the u and c quarks have the same mass. Of course, the cancellation could still be arranged by varying coupling constants to suit, but it would then be ad hoc and contrived. Consequently, those who propose the charmed quark c generally suggest that the lightest charmed hadrons should have masses not more than several GeV - say 5 GeV at the very outside. How would charmed hadrons be seen? The charmed hadrons would undergo the usual hadronic and electromagnetic decay modes, until the sequence reaches some state which is the lightest hadronic state for given B, s, Q and C. Such states can only undergo weak decay, either semi-leptonic, to ($\ell+\nu_\ell$+C=0 hadrons), or hadronic, to (C=0 hadrons). One feature of the simple GIM scheme is that the dominant $\Delta C=-1$ term in the hadronic weak current formed from the doublet (16) is $(\bar{c}s)\cos\theta_C$, which gives the transition $c\to s$, i.e. $\Delta s=-1$. Hence the GIM scheme predicts that the semi-leptonic decays of charmed hadrons should be dominated by final states with strangeness. Since strange particles are not among the products for many of the weak decay processes known, this appears a promising signature in the search to identify these charmed particles.

At a recent London Conference, Iliopoulos(16) was so much enthused about the GIM scheme that he asserted that charmed particles are being produced in accelerator experiments now and that experimenters have simply not looked hard enough, or not in the right way, to see them. Further, he offered to all comers an open bet of a case of wine that charmed particles will have become established before the next Conference in this series. This was an extremely rash offer! Just remember that it is only two years until the next Conference, to be held in Russia in 1976 - also, that most experiments in high energy physics take four to five years from conception to final results. However, this offer is a good measure of the strength of one theoretician's belief and enthusiasm concerning this proposed quantum number "charm".

Our discussion of this topic could go on for much longer, but my time is up. I apologize to any in the audience who may have their pet "heavy particle" which I have not mentioned. There are certainly many other possibilities - heavy leptons, the Higgs particles, and so on. In every case, I have confined my remarks to the simplest possible version of the schemes proposed - all of them can be extended and complicated in an infinity of ways, although there are generally sub-classes of these extensions which have their own special attractiveness. In conclusion, I should like to say that I feel happy in believing that, with the machines which we can see on the immediate horizons or those most accessible in the next generation of devices, the questions which we have exposed here should be settled well before the time comes for my own retirement.

Note added (March 1976). In order to see the above remarks in a proper perspective today, it seems desirable to add a brief account of the developments which have taken place in the twenty months which have passed since this lecture was given.

In November 1974, there occurred the dramatic discovery of the J/ψ vector meson of mass 3095 MeV, reported from two quite different investigations, one experiment being carried out with the proton beam from the AGS at Brookhaven National Laboratory(17) and the other with the electron-positron storage ring SPEAR at the Stanford Linear Accelerator Center.(18) Observations of two further vector mesons, $\psi'=\psi(3684)$ and $\psi(\sim 4100)$, were reported from the SPEAR experiment during 1975,(19) and one further vector particle $\psi(4414)$ has become established quite recently.(20) It appears now that there may be appreciable structure yet to be resolved, within the broad $\psi(\sim 4100)$ peak.(21) A careful search for narrow resonances has been made up to 5.9 GeV, but without further discoveries.(22) The first two of these particles, $J/\psi(3095)$ and $\psi(3684)$, are notable for their astonishingly narrow decay widths, 0.07 MeV and 0.23 MeV, respectively, but the upper states are broad, the widths being about 200 MeV for $\psi(4100)$ and 33 MeV for $\psi(4414)$.

In terms of the ideas mentioned in the text above, the most immediate explanation of these new heavy meson states was that they were related with the fourth quark c. In this view, $J/\psi(3095)$ was the charm analogue of the vector meson $\phi(1020)$, the two mesons being the lowest $^3S_1$ states for the $\bar{c}c$ and $\bar{s}s$ systems, respectively. For the meson ψ', the fact that its dominant decay mode was a transition to J/ψ with the s-wave emission of an s-wave $\pi^+\pi^-$ pair, suggested that it had a similar structure to J/ψ, and it was therefore assigned to the radially excited configuration 2^3S_1. The higher states, $\psi(4100)$ and $\psi(4414)$, then receive a natural interpretation as the n=3 and n=4 states of the $^3S_1$ configuration. With the simple assumption that the $\bar{c}c$ potential has a linear radial dependence, as proposed on the basis of quark confinement schemes, this empirical sequence of mass values fits quite well with expectation for the principal series of states n^3S_1, when the parameters are adjusted to fit the mass values for J/ψ and ψ'. This interpretation requires that the c quark should have a significantly larger mass than the s quark, by about 1 GeV, a larger difference than that between the s quark and the (u,d) quarks, various estimates of the latter difference giving values between 150 and 300 MeV.

The extraordinarily narrow decay widths for J/ψ and ψ' are reminiscent of the unusual narrowness of the $\phi(1020)$ meson, whose decay width is $\Gamma(\phi)=4.2\pm0.2$ MeV. Of course, $\phi(1020)$ is above the $K\bar{K}$ threshold, so that the decays

$$(\bar{s}s) \to ((\bar{s}d) + (\bar{d}s)) \text{ and } ((\bar{s}u) + (\bar{u}s)) \qquad \text{(A1)}$$

are energetically possible. However, the decay $\phi \to \pi^+\pi^-\pi^o$, which involves an annihilation of the initial s and $\bar{s}$ quarks, has a partial decay width of only 0.65 MeV. This has always been interpreted in terms of the Zweig rule that all quark line graphs which involve annihilation of a quark and an antiquark which belong to the same initial or final particles should be omitted. This rule actually forbids the decay $(\bar{s}s) \to \pi^+\pi^-\pi^o$, and the observed ϕ decay is attributed to a small admixture of $(\bar{u}u+\bar{d}d)$ state to the $(\bar{s}s)$ state, in the physical ϕ meson state. The application of the Zweig rule to hadronic decay modes for J/ψ and ψ' proceeds in complete analogy to the foregoing discussion for $\phi(1020)$ decay to pions. The decay processes

$$(\bar{c}c) \to ((\bar{c}d) + (\bar{d}c)),\ ((\bar{c}u) + (\bar{u}c)) \text{ and } ((\bar{c}s) + \bar{s}c)) \qquad \text{(A2)}$$

are apparently forbidden for J/ψ and ψ' because they lie beneath the thresholds for these final states (see below), and the observed decay processes are only those which are possible in virtue of admixtures in the physical J/ψ and ψ' states, or of the annihilation process $(\bar{c}c) \to$ "γ" $\to$ hadrons. What has appeared so remarkable is the high degree of $(\bar{c}c)$ purity which the J/ψ and ψ' states must have in order to account for such narrow decay widths for such heavy particles.

The simple $(\bar{c}c)$ potential model mentioned above also predicts that there should exist three ^3P_J states for this system, intermediate between the J/ψ and the ψ' masses, with spin-parity values (0+), (1+) and (2+). Several authors predicted that there should be found a sequence of γ transitions passing through these states, as follows:[23,24]

$$\psi' \to \gamma + \psi(^3P_J), \qquad \text{(A3a)}$$

$$\psi(^3P_J) \to \gamma + \psi. \qquad \text{(A3b)}$$

There is now clear evidence[25,26] for two (perhaps three) transitions of the type (A3a), and one of these transitions is followed by a transition of type (A3b); apparently the state $\psi(^3P_J)$ at 3500 MeV is such that the γ transition to ψ can compete successfully with its hadronic decay modes. This agreement, in broad outline at least, is reassuring as concerns the correctness of the

$(\bar{c}c)$ model for the ψ meson family.

The large widths observed for $\psi(\sim 4100)$ and $\psi(4414)$ are readily understood if these states lie above the lowest of the thresholds (A2), but what are the mass values expected for the C=+1 charmed mesonic states specified there, $D^+=(\bar{d}c)$, $D^o=(\bar{u}c)$ and $S=(\bar{s}c)$? If we follow the prescriptions which have proved effective for the corresponding situation concerning the s quark, we are led to the mass estimate 2260 MeV for the vector meson D_V and 2300 MeV for the vector meson S_V. The $^1S_0$ states for the $(\bar{c}c)$ system are expected to lie somewhat below these values. Adopting again the prescriptions successful for the s-quark situation, we are led to the mass estimates 2130 MeV for D_{PS} and 2180 MeV for S_{PS}. These are only rough estimates, but they are sufficient to make it quite plausible that the threshold for charmed meson pair production may well lie at about 4 GeV and so account for the large widths observed for the ψ mesons $\psi(\sim 4100)$ and $\psi(4414)$.

We must emphasize here that no evidence has yet been found in the final hadronic states studied in detail in the SPEAR experiments to demonstrate that the pair creation process $e^+e^- \to D^+\bar{D}^-$ does actually occur, even at total c.m. energy 4.8 GeV (where there is a large body of data). It seems possible that the D meson decay processes may be dominantly hadronic, with many multiparticle final states, which would be difficult to identify. Experimental studies have been made with high statistics for all final states $K\pi$ and $K\pi\pi$, both at SPEAR and in hadron reaction processes at other accelerators, without any indications for a charmed meson. We now have the bizarre situation that there is a strong indirect case for the introduction of the charm quantum number C, although we have not yet identified empirically any hadronic state which has non-zero value for C. If Iliopoulos's bets are still on, there is little time left for his claim to be proved valid, since the High Energy Conference at Tiflis is less than four months off.

Another important aspect of the SPEAR experiments has been reported by Perl et al.[27] They have studied those events in which the final state includes an electron and a muon of opposite charges and no other charged particles (showers resulting from π^o may be seen, or $K^o_s \to \pi^+\pi^-$ decay when this occurs after the charged particle veto). Within this data, they have identified a substantial class of events of the type $(\mu^{\pm} + e^{\mp}$ + unobserved neutrals) in which the electron and muon are not generally collinear. In view of the above discussion, the hypothesis arises naturally that these events may be due to the $e^+e^- \to D^+\bar{D}^-$ process, with the subsequent decay modes $D^+ \to \mu^+\nu$, or $\mu^+\nu(\pi^o$ or $K^o)$, etc., and $D\bar{D}^- \to e^-\bar{\nu}$, or $e^-\bar{\nu}(\pi^o$ or $K^o)$ etc., or vice versa for μ

and e. This interpretation is ruled out on two distinct grounds,
(a) that the rate observed for $\mu^{\pm}e^{\mp}$ events accompanied by some number of π^{o} or K_{s}^{o} particles which are detected is too low, unless the modes $D^{+} \rightarrow \mu^{+}\nu$ and $e^{+}\nu$ were the dominant modes - but this is ruled out by the low rate found for ($\mu^{\pm}e^{\mp}$ + unobserved neutrals),
(b) the spectrum of leptons observed in the c.m. frame does not agree with two-body production $D^{+}\bar{D}^{-}$ followed by the two-body decay modes $D^{+} \rightarrow \ell^{+}\nu$, $\bar{D}^{-} \rightarrow \ell^{-}\bar{\nu}$, but it does fit quite well if a three-body decay mode is assumed - but the three-body modes $D \rightarrow (\ell\nu(\pi^{o}$ or $K_{s}^{o}))$ are already ruled out by (a). Perl et al. therefore conclude that what they are seeing is

$$e^{+}e^{-} \rightarrow U^{+}U^{-}, \tag{A4}$$

for some particle U with mass between 1.8 and 2.0 GeV, which has the decay mode

$$U^{+} \rightarrow \ell^{+} + \text{(at least two neutral, unobserved particles)}. \tag{A5}$$

The most direct interpretation of these observations reported by Perl et al. is that the particle U is a heavy lepton, with the decay mode

$$U^{+} \rightarrow \ell^{+} + \nu_{\ell} + \bar{\nu}_{U}, \tag{A6}$$

where ν_{U} denotes the neutrino associated with the lepton U. The magnitude of the cross section and the $\mu^{\pm}e^{\mp}$ event rates fit the data if the U lepton has spin $\frac{1}{2}$ and the branching ratios for the decay modes (A6) are 17% for final muon and 17% for final electron. The other decay modes possible are of the type

$$U^{+} \rightarrow \bar{\nu}_{U} + \text{hadrons}. \tag{A7}$$

The hypothesis of the existence of a series of heavy leptons was much discussed a few years ago, in connection with gauge theories of the weak currents and the unification of weak and electromagnetic interactions. There were two motivations. The first was that, if heavy leptons were added, this unification could be achieved just with the charged bosons $W^{\pm}$, and in such a way that the only neutral current involved was the electromagnetic current; the prototype for such schemes was that of Georgi and Glashow,(28) who added two sets of heavy leptons, an electron type (E^{o}, E^{+}) and a muon type (M^{o}, M^{+}). The second aim was to unify the electron and muon into a single multiplet, to obtain perhaps some understanding of the large muon/electron mass ratio and to determine what constraints might follow concerning the properties of heavy leptons. The possible schemes were surveyed then by Bjorken and Llewellyn Smith.(29) These early discussions were not followed up because of the lack of empirical evidence for heavy leptons and because gauge theories based on the charm quantum number already were sufficient to achieve the unification of the weak and electromagnetic interactions while still satisfying the empirical constraint that there be no $\Delta s=\pm 1$, $\Delta Q=0$ weak currents.

The U-pair production process (A4) gives a cross section which rises rapidly from the threshold energy $2M_U$. However, this process cannot account for all of the increase seen for the total cross section $\sigma(e^+e^-)$ between (say) c.m. mass 3750 and 4250 MeV, nor can it account for the large widths observed for $\psi(\sim 4100)$ and $\psi(4414)$. Hence we still require the existence of charmed mesons D with threshold mass $2m(D) \sim 4$ GeV, in order to account for these striking features of the data. It is certainly curious that the threshold for the states $U\bar{U}$ and $D\bar{D}$ should lie so close in mass. It suggests that there may be some relationship between the lepton U and the meson D, but none is yet known.

References

1. N. Byers and R. E. Peierls, Nuovo Cim.10 (1958) 520.
2. A. Salam and J. C. Ward, Phys.Lett.13 (1964) 168.
3. S. Weinberg, Phys.Rev.Lett.19 (1967) 1264.
4. This is the value currently obtained from the studies of inclusive hadronic neutral currents with Gargamelle at CERN. The values obtained from the high energy experiments at Fermilab and from the study of leptonic neutral currents are compatible with this value but with larger uncertainty. (Private communication from G. Myatt, Oxford University).
5. G. t'Hooft, Nucl.Phys.B33 (1971) 173 and B35 (1971) 167.
6. D. H. Perkins, Contemp.Phys.16 (1975) 173.
7. V. F. Weisskopf, "The MIT Bag 1975", CERN preprint TH.2068 (1 September 1975).
8. M. Gell-Mann, in Elementary Particle Physics (ed. P. Urban, Springer-Verlag, Vienna, 1972) p.145.
9. H. J. Lipkin, Phys.Lett.45B (1973) 267.
10. A. De Rujula, G. Georgi and S. L. Glashow, Phys.Rev.D (1975) 147.
11. G. Feldman and P. T. Matthews, Phys.Rev.Lett.35 (1975) 344; Nuovo Cim.31A (1976) 447.
12. T. Massam and A. Zichichi, Lett. Nuovo Cim. 1 (1969) 388.
13. M. S. Chanowitz and S. Drell, Phys.Rev.D9(1974) 2078.
14. G. B. West, Phys.Rev.D10 (1974) 329.
15. S. L. Glashow, J. Iliopoulos and L. Maiani, Phys.Rev.D2 (1970) 1285.
16. J. Iliopoulos, in Proc.XVII Intl. Conf. on High Energy Physics (ed. J. R. Smith, Rutherford Laboratory, Science Research Council, Chilton, Didcot, U.K., 1974), p.III-89. See especially p.III-100.
17. J. Aubert et al., Phys.Rev.Lett.33 (1974) 1624.
18. J. Augustin et al., Phys.Rev.Lett.33 (1974) 1406.

19. Particle Data Group, Review of Particle Properties: Supplement to 1974 Edition, Revs.Modern Phys.47 (1975) 535.

20. J. Siegrist et al., Phys.Rev.Lett.36 (1976) 700.

21. B. Richter, "SPEAR Results on the Total Cross Section, New Particles and μ-e Events", invited paper at the Royal Society Discussion Meeting on "New Particles and New Quantum Numbers", held in London on 11 March 1976, Proc.Roy.Soc.A., to be published (1976).

22. A. M. Boyarski et al., Phys.Rev.Lett.34 (1975) 762.

23. E. Eichten et al., Phys.Rev.Lett.34 (1975) 369.

24. T. Appelquist, A. De Rujula, H. Politzer and S. L. Glashow, Phys.Rev. Lett.34 (1975) 365.

25. G. J. Feldman, in Proc. Intl. Symp. on Lepton and Photon Interactions at High Energies (ed. W. T. Kirck, SLAC, Stanford Univ., California, 1976) p.39.

26. B. H. Wiik, ibid., p.69; H. Heintze, ibid., p.97.

27. M. Perl et al., Phys.Rev.Lett.35 (1975) 1489. See also M. Perl, "Properties of eμ Events produced in e^+-e^- Annihilation", Rept.SLAC-PUB-1664, November 1975.

28. H. Georgi and S. L. Glashow, Phys.Rev.Lett.28 (1972) 1494.

29. J. D. Bjorken and C. H. Llewellyn Smith, Phys.Rev.D7 (1973) 887.

Discussion

Weisskopf: I want to ask a question about the presence of antiquarks in the nucleon. You gave a number. First, I would like to know how you define this. Are the antiquarks weighted according to the Feynman x or not? In other words, for example, could the number be understood in terms of many antiquarks, but all moving slowly, that is, with low values of x?

Dalitz: Yes, I believe that you are right, the distribution is weighted by a factor of x.

Weisskopf: You see, I am troubled by the fact that after all we know that there is a reasonably good probability that a nucleon is part of the time another nucleon plus a pion cloud, and the pion cloud is supposed to have 50% antiquarks in it. It is true that, if I pursue this line of thought, and look at it the Feynman way, with the nucleon moving very fast, the momentum fraction carried by the pion cloud is very small, about 1/7, which has to be divided again by two because it is shared between quarks and antiquarks, so that you get a rather small x; but still, there ought to be a number of antiquarks there, which correspond to the pion cloud, and this is the point that worries me.

Dalitz: Yes. There are direct experiments that measure the x distribution for antiquarks, and you can compare that with the x distribution for quarks in the nucleon. The antiquarks are indeed concentrated at low x. However, since their number appears with the factor x in cross sections, they play a relatively small role in these sum rules.

Peierls: Could I try to clear up what seems to me a misunderstanding here. If you start without any interactions, putting in three quarks corresponding to the sum rule that Dalitz mentioned, you could then put in all sorts of virtual processes like the ones you have in mind, but for every antiquark you added you would also have to add a corresponding quark, and so these processes would not show up in that sum rule, which actually measures the difference between the numbers of quarks and antiquarks.

Weisskopf: I am sorry. Maybe I did not express myself clearly. I am aware that whenever you have a pion you have to have not only antiquarks, but also quarks, but the point is that the present experiments give the impression that there are no antiquarks. Now of course, there is always an accuracy limit, and this is what my question implied - is this limit in contradiction with that primitive picture that I painted, or is it not? I believe it is not, but I am not quite sure.

Peierls: You are talking about experiments, and not the sum rule?

Weisskopf: Yes, there is, for example, the neutrino-antineutrino ratio which is a very sensitive indication; a less reliable indicator are the so-called Ledermann experiments, where you produce muon pairs by the collisions of two protons, and there should be a quark/antiquark annihilation. Now, that is not on such safe ground. But all these seem to indicate there are very, very few antiquarks, at least of reasonable momentum, and there is a point that ought to be cleared up.

Dalitz: Yes, the neutrino experiments do suggest that there are fewer antiquarks in the nucleon than are indicated by the over-all fit of this model to all the data, but the discrepancy is less than two standard deviations at present.

Karl: I wish to comment on the possibility that these particles never exist in free space, and would like just to make the historical parallel to the case of beta decay, when apparently people - some of whom are probably here - were unhappy at the time at the possibility that electrons came out of nuclei, while they were not there inside in the first place. Apparently, there was a certain amount of discussion in the 1930's, when the neutron-proton model for the nucleus was put forward, and people felt unhappy that particles came out that were not there in the first place. The situation at the moment is exactly the reverse of that. You are wondering whether you could have particles there that could never come out, and at least from a psychological point of view, I think one feels slightly better by remembering that people were equally unhappy by the opposite situation!

8

An Old-Timer Looks at Modern Field Theory

V. F. WEISSKOPF

I have chosen this topic for my talk because I was thinking about the old days when I first met Rudi in the year 1933, when we were both Rockfeller fellows. I was extremely glad to have received this fellowship because, as you may still remember, 1933 was a period when it was hard to find jobs - not only for physicists - in fact much harder than now, which I never dare to tell present-day young people. I remember our stipend still, I got $150.00 per month, but Rudi got $200.00 per month, and I was of course very curious to know why he was favoured. The justification was that he was married at that time; if I'd only have known! I also remember that we were always obliged to write to the Rockefeller Administration whenever we had made an "achievement", whatever that means. I remember that Rudi wrote a letter when Lady Peierls gave birth to Ronnie. I am sure that this was not well received at the office in Washington. How could they have known about the future services of Ronnie to U.S. science!

The reason why all this has to do with field theory, is that I learned from Rudi most of what I know in this field. I am younger than Rudi - not by much - but it is interesting to observe that this Δy - the difference in age - plays a very big rôle when you are young, and I am afraid it begins also to play a very big rôle when you are old. At that time, I definitely was younger than Rudi, and I was very glad to have somebody who could teach me all the tricks one uses when one applied this wonderful field theory, that was conceived a few years before by Dirac. He showed me how to solve a lot of electron problems, such as positron theory, pair creation, etc., and I remember that I learnt all that - it was called at that time "alpha gymnastics" - from Rudi at his house when he had to help Genia to care for the baby. He not only taught me about alpha gymnastics. It was at that time when I got an offer to become the assistant of Pauli, and he also taught me very successfully how to deal with Pauli - but this is another matter.

Let us turn to our subject: field theory is a very important subject because it is the only way we physicists can deal in some consistent manner with the interaction between particles; we do not know whether it is the right way, but it is the only way we know; it is the only language in which we can speak about interactions. Whether nature speaks the same language is a very different question. I would say it is not very probable. At best, the difference between nature's language and our language is comparable to the difference between the English you speak in England, and the English they speak in the middle-west of America. But I suppose the difference is much deeper than that - deep as this difference is. Now, another disadvantage of field theory is that it is in some ways not a completely logical language, well, that it has in common with all languages. There are still infinities in the theory, but we now know how to deal with it in sort of a fashion - the so-called renormalization procedures. They represent - as S. Drell has put it - not a real solution, but only "peaceful coexistence" with the infinities.

What I want to tell you about today is something different. I always tried, and Rudi also, to get some kind of a simpler way to see what happens in this rather abstruse mathematics that we are usually encountering in field theory. I am doing this because, at present, field theory becomes again a very important tool and it is enlarged. We have heard from Dalitz about the Weinberg theories, but there are also other very important points which I would like to mention and so I follow an old, but not too old, tradition to look at these things in a simple, physical, "hand-waving" way. In contrast to most people, I do not consider hand-waving to be necessarily a deprecatory adjective.

I would like to present* four examples of intuitive reasoning in field theory. The first three are concerned with problems of quantum electrodynamics, namely, the Lamb shift, the radiative correction to the magnetic moment of the electron (the deviation of the electronic g factor from 2), and

* Most of this material was presented to the Erice Summer School 1974, and is published in their proceedings.

the polarization of the vacuum by an external charge. The polarization of the vacuum is a simple example of a typical fact resulting from quantum electrodynamics: the effective charge e' for processes in which momentum transfers $q \gg m$ (electron mass) occur, increases with larger q as $e' \sim \log (q/m)$. The fourth example deals with an intuitive approach to the problem of "asymptotic freedom", a term used for the remarkable fact that in certain field theories the effective charge e' decreases with larger g as $e' \sim (\log q/m)^{-n}$ where n in the simplest case is $\frac{1}{2}$. These field theories are usually referred to as "non-Abelian". We will not comment on that term, but restrict ourselves to the remark that in these field theories not only the particles but also the fields are carriers of charge.

1. Lamb Shift

An intuitive understanding of the Lamb shift is based upon the recognition that there is an influence of the zero-point oscillations of the electromagnetic field upon the motion of the electron (1). The intensity of the field oscillations can be expressed in terms of the average square of the electric field fluctuations in the vacuum which, per frequency interval $d\omega$, are given by

$$\mathcal{E}^2_{fl} = \omega^3 d\omega \tag{8.1}$$

In all relations of this article, the equal sign is meant to signify equality apart from numerical factors. A free electron performs forced oscillations in that field whose amplitudes Δx are given by the classical non-relativistic relation, as long as $\omega < m$:

$$(\Delta x)^2 = e^2 \mathcal{E}^2_{fl}/m^2\omega^4 \quad \text{for } \omega < m. \tag{8.2}$$

When the frequency is large compared to m, quantum field theory (the effect of the positron field) shows that the classical expression (Eq. 8.2) is reduced by $(m/\omega)^2$:

$$(\Delta x)^2 = e^2 \mathcal{E}^2_{fl}/\omega^6 \quad \text{for } \omega > m. \tag{8.3}$$

Inserting Eq. 8.1 into Eq. 8.2 and Eq. 8.3 and integrating over all frequencies gives the average total amplitude. It converges at high frequencies because Eq. 8.3 acts like an upper cut-off at $\omega = m$. When considering a bound electron in the hydrogen atom, a cut-off at low frequencies appears since the electron is no longer free for $\omega < \omega_o$, where ω_o is of the order of a Rydberg: $\omega_o \sim Ry$. So we get

$$(\Delta x)^2 = \frac{e^2}{m^2} \log \frac{fm}{\omega_o} \tag{8.4}$$

where f is a numerical factor.

Thus the position of the electron fluctuates by Δx. This introduces a change in kinetic and potential energy. The former change is the same (in that order of e^2) for a free and a bound electron; it is therefore part of the renormalized mass. The change in potential energy ΔE_L is the Lamb shift. Call $V(x) = -e^2/|x|$ the potential energy in the H-atom, and $|\psi(x)|^2$ the probability of finding the electron at x; then we get

$$\Delta E_L = \int dx^3 |\psi(x)|^2 (\Delta x)^2 . \nabla^2 V = |\psi(o)|^2 e^2 (\Delta x)^2 ,$$

since the Laplacian of V is a delta function at the origin. We now use Eq. 8.4 and the relations

$$|\psi(o)|^2 = \frac{1}{a_o^3} , \frac{e^2}{a_o} = Ry, (ma_o)^{-1} = e^2$$

where a_o is the Bohr radius, and we get

$$\Delta E_L = Ry\ e^6 \log \frac{fm}{\omega_o} \tag{8.5}$$

for the Lamb shift. The natural line width is of the order e^6Ry; thus the Lamb shift is larger only by a logarithmic factor.

2. (g - 2) Considerations

Let us now consider the radiative corrections to the magnetic moment μ of the electron. Here also it is the effect of the vacuum field fluctuations which causes a deviation from the Dirac value $\mu = e/2m$.

There are two effects at work: one reduces the magnetic moment, the other increases it. The second effect is larger. The first effect is an exchange of angular momentum between the electron and the radiation field. An electron with its spin up emits a virtual quantum by flipping its spin, and reabsorbs it again. This will lead to a decrease of the effective magnetic moment, since the electron spin is part of the time reversed. The fraction of the time being in the opposite direction must be proportional to e^2. Hence the correction $\Delta\mu_I$ of the first effect is approximately

$$\Delta\mu_I = -e^2\mu .$$

Another way of describing this effect goes as follows: the spin of the electron suffers fluctuations of direction due to the influence of the vacuum fluctuations of the magnetic field. Let us use the equation of motion of the spin:

$$\dot{\sigma} = \mu_o (B \times \sigma)$$

where μ_o is the magnetic moment of the electron, $\mu_o = e/2m$. For a given frequency interval around ω, we find for the change $\Delta\sigma$

$$\frac{\Delta\sigma}{\Delta t} = \Delta\sigma.\omega \sim \frac{e}{2m} B_{\perp}.\sigma$$

where $B_{\perp}$ is the component of B perpendicular to σ. We now determine the average angle $\theta = \Delta\sigma/\sigma$ by which the spin fluctuates around its direction:

$$\theta^2 \sim \frac{e^2}{4m^2\omega^2} B_{\perp}^2.$$

Using for B^2 the same estimate as for the electric fluctuations in Eq. 8.1, and integrating over $d\omega$, we obtain

$$\theta^2 \sim \frac{e^2}{4m^2}\int_o^m \omega d\omega \sim \frac{e^2}{8}$$

Here the upper limit comes from the fact that the magnetization, coming from circulating currents, is spread over an area of linear dimensions m^{-1}. The reduction of the magnetic moment caused by these fluctuations is, of course,

$$\Delta\mu = - \frac{\theta^2}{2} \mu_o \sim - \frac{e^2}{16} \mu_o. \qquad (8.6)$$

The second effect* is based upon the influence of the vacuum field fluctuations on the circular current which produces the magnetic moment. The expression for the electric current of the relativistic electron can be split into a translational and a circular current (Gordon split). The latter one corresponds to a loop of radius $1/m$, around which the electron circles with light velocity, that is, around which flows a current $I = em$. The magnetic moment of a circular current with radius r is

$$\mu = \frac{1}{2} Ir^2$$

giving just the correct value of the magnetic moment. The fluctuations Δx of the position of the electron induced by the vacuum-field oscillations, produce fluctuations of the radius r of the loop around its undisturbed value m^{-1}. Hence we may write:

---

* The first effect was already mentioned in T. D. Welton (1). The second one was suggested by Z. Koba (2).

$$r^2 = \frac{1}{m^2} + (\Delta r)^2$$

and this correction gives rise to a change of the magnetic moment:

$$\frac{\Delta\mu_{II}}{\mu} = \frac{(\Delta r)^2}{r^2} . \tag{8.7}$$

We insert for $(\Delta r)^2$ an expression similar to the one we used in the calculation of the Lamb shift. But there is an important difference. The wavelength of the fluctuating field must not be larger than the radius $r = 1/m$ of the loop; if it were, the loop would be displaced as a whole and the radius would remain unchanged. Hence only frequencies $\omega \geq m$ are effective for $(\Delta r)^2$. This fact makes the logarithm contained in Eq. 8.4 disappear and we get simply:

$$(\Delta r)^2 = \frac{e^2}{m^2} \tag{8.4.1}$$

and therefore $(\Delta r)^2/r^2 = e^2$ which leads to a positive change of the magnetic moment (see Eq. 8.7)

$$\frac{\Delta\mu_{II}}{\mu} = e^2. \tag{8.8}$$

These considerations do not directly show that $\Delta\mu_{II}$ from Eq. 8.8 is larger than the absolute value of $\Delta\mu$ in Eq. 8.6, in order to guarantee that the overall $\Delta\mu$ is positive. Koba has shown that it is true. As a vague indication, we mention that Eq. 8.6 contains a rather small numerical constant.

3. Vacuum Polarization

The attraction between two opposite charges e and -e is not exactly given by the law of Coulomb; there is a correction at distances $R \leq 1/m$ which is determined by the so-called Uhling function U(R). The potential energy of the two charges is

$$E = -\frac{g^2(R)}{R}$$

where g(R) is the effective charge at the distance R. We find

$$g^2(R) = e^2(1 + U(R)). \tag{8.9}$$

The form of U(R) is simple if R is small or large compared to 1/m:

$$U(R) \to \frac{2}{3} \log mR \text{ for } R << m^{-1}$$

$$\to \frac{\pi^{1/4}}{4} (mR)^{3/2} e^{-2mR} \text{ for } R >> m^{-1}. \tag{8.10}$$

The effective charge increases logarithmically tow-ds smaller distance if

$R < m^{-1}$.

The interpretation of this deviation from Coulomb law is as follows. In order to avoid infinities, let us introduce a cut-off at some very small distance R_o and let us put a charge $e_o > e$ at the centre within R_o. This charge induces a polarization of the vacuum proportional to the field strength. This polarization will be such that the vector of polarization points in the positive R direction if the charge in the centre is positive. This must be so, since the virtual pairs will arrange themselves such that the positive charge is pushed away from the centre. The strength of the polarization obviously diminishes with increasing R, and therefore an induced charge ρ' is produced whose sign is opposite to the sign of the centre charge. In first approximation this charge is

$$\rho' = -Ce^3/R^3 \qquad R \ll m^{-1}$$

$$\rho' \to 0 \qquad R \gg m^{-1}.$$

This expression follows from the fact that, for $R < m^{-1}$, the polarization is proportional to the field strength e/R^2 with a proportionality factor e^2, and the charge density is the divergence of the polarization. The constant C is such that the integral over ρ' equals $e - e_o$. Then the net charge seen at distances $R > m^{-1}$ is just e. Thus in quantum electrodynamics, the higher the momentum transfer (small distances), the nearer one gets to the bare charge which is larger than the normal one.

4. Asymptotic Freedom

Recently the theoretical observation was made that certain field theories differ fundamentally from quantum electrodynamics, in that the coupling becomes weaker and not stronger with increasing momentum transfer Δ_q and goes to zero with the logarithm of Δ_q. It is aptly called asymptotic freedom and occurs in so-called "non-Abelian" theories. The important physical character of these theories is the fact that the field itself carries charge, and therefore acts as a source of fields. We will show that a decoupling at small distances is indeed expected under these conditions. At the same time it will emerge that the description of phenomena at large distances (small momentum transfer) will become much more intricate and difficult compared to field theories with neutral fields such as quantum electrodynamics.

Let us discuss a simple extension of quantum electrodynamics in which the field carried charge. We consider the field vector $\vec{A}$ as a vector in an isospin space, and the particle spinor a spinor in that space, and write the interaction H' in the form

$$H' = e(\psi\gamma_\mu\vec{\tau}\psi)\vec{A}_\mu. \tag{8.11}$$

This interaction allows for a charge transfer from the field ψ to the field A and vice versa. We now discuss the effect of this generalization upon the vacuum polarization.

We find two opposite effects. On the one hand, the "bare" charge e_o (let us assume e_o is positive), as defined in the previous section, will produce a vacuum polarization of the same type as discussed before. This will induce a negative charge density ρ' in the neighbourhood ($R < m^{-1}$) of e_o. On the other hand, the interaction of Eq. 8.11 also produces a charge exchange between the source e_o and the field surrounding it. This effect produces a positive charge density in the neighbourhood of e_o. The two effects appear in a perturbation approach in the same order (e^3), as Figs. 8.1(a) and 8.1(b) indicate.*

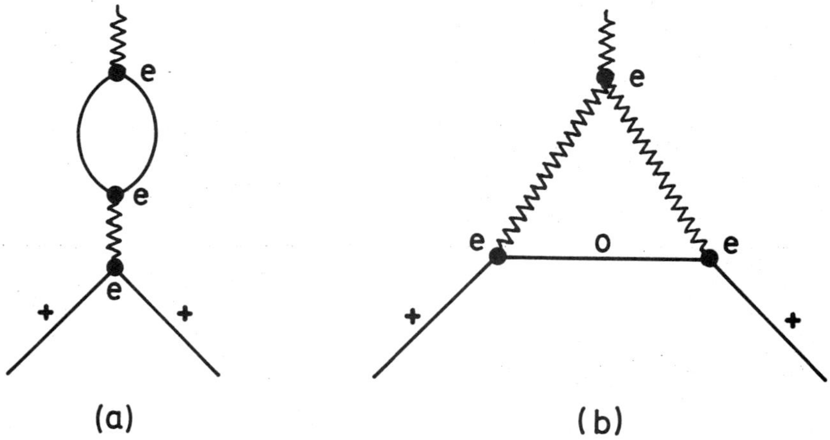

Fig. 8.1. (a) Vacuum polarization. (b) Charge exchange with field.

We cannot show in a simple way which of the two effects is the stronger. The calculations indicate that the positive charge density wins, except in cases

* It should be added that the division of the effect into those two diagrams is not gauge invariant. Nevertheless, the division is useful; in some gauges the physics is more directly evident than in others.

where there are a large number of different ψ fields (an understandable effect). Thus, with charge carrying fields, the induced charge surrounding the bare charge has the same sign as the latter. Since the total charge is fixed, the bare charge must vanish at the centre; the theory is asymptotically free.

A theory with charge transfer to the field will encounter difficulties at low momentum transfers. This is best seen when contemplating how quantum electrodynamics avoids the difficulties coming from the fact that the probabilities of emitting and absorbing very low frequency quanta become infinitely large. One describes the long wavelength fields classically (Bloch-Nordsieck method); the classical description becomes most appropriate at low ω, because the quantum fluctuations become less important and can be neglected, since the number of quanta contained in the classical field is growing large. However, when there is charge exchange between source and field, the situation is different. The charge of the field is either unity or zero. Thus the quantum fluctuations can no longer be neglected: the fluctuations of the charge are of the order of the charge itself. It therefore is not possible to use the classical field as a classical limit at low ω.

This is why the infra-red situation of charge transfer theories is yet to be understood. It is possible that unexpected features may turn up which may help to describe some of the phenomena seen in strong interactions, such as the non-existence of free quarks.

In this connection, it may be of interest to present an idea that I learned from Sidney Coleman, indicating that the vacuum is not stable, in some sense, in such a theory. Why? Take two field quanta, one + and the other -, not fermions but Yang-Mills particles, with mass zero. The potential energy between two such charges is less than $1/r$, or in absolute amount more than $1/r$. That follows from the theory, but one can also easily understand why, because the field between the particles is charged and therefore the charges are brought effectively nearer. One also can say that the field lines of the two charges attract each other, thus increasing the force. Therefore, one will get a force expression which comes from a potential that is stronger than $-1/r$; I am told, though I have not calculated it, that it is $(-1/r)\,\ln r$. The quantum kinetic energy of two particles is not strong enough to balance it because, for two massless particles at a distance r, it is $1/r$. So we get the feeling that such a situation of two oppositely charged particles has an energy less than the vacuum. And therefore the vacuum may not be stable.

This is a negative statement, and it is meant as a negative statement. As you know, the theorists have not really been able to answer what happens in non-Abelian theories in the infra-red limit, and they get into one trouble after the other. They hope - and this is a hope - that maybe such a theory will give what is another feature of the strong interaction apart from the asymptotic freedom: namely the feature that the quarks are kept in a "bag" - that the quarks are kept together by that force, they cannot be free. Now, it may be that this kind of "phase-transition" in the vacuum may lead to a situation in which indeed these theories do give the result that the potential energy between two quarks goes to infinity when the distance between them becomes large. I do not know - I am not even necessarily optimistic about it.

Well, this is in some ways all I want to say. There are, of course, more new ideas in modern field theory than that - Dick Dalitz told you about the wonderful ideas about the unification of weak and electromagnetic interactions. I wish I had some hand-waving arguments also to make Weinberg's theory a little clearer - I do not have - (also my time is running out) - but I think one should have. Altogether, and here again I speak as an "old-timer", I feel that the purely mathematical approach is important, but it is not the only one. Quite apart from the fun and insight one gets when one thinks very hard about what some people call hand-waving (and I would rather call qualitative physics) arguments, I also believe that such arguments are most helpful to find out in what direction one should go for the solution of some of the outstanding problems.

To end my talk on a more humorous note, I would like to quote to you an anecdote, in which two people from two different Jewish villages compare their Rabbis. One says, "Well, our Rabbi is really a wonderful Rabbi because everything he says is true, and he never says anything that is not true." And the other says, "Oh, our Rabbi is much better; surely everything he says is true, but he also tells us what is really true."

References

(1) T. Welton, Phys. Rev. 74 (1948) 1157.

(2) Z. Koba, Prog. Theor. Phys. 4 (1949) 319.

CLOSING REMARKS

That brings us astonishingly quickly and very pleasantly to the end of this Symposium in honour of Professor Peierls, and I'm sure you wouldn't like me to declare it closed without a few appropriate words. Our most important thanks go to Sir Rudi: for being himself, and for bringing us all here over tremendous distances, and also tremendous stretches of time. Everybody here owes Rudi a great deal, many of us in many different ways. And, you know, the keynote of the Symposium has not been reverence for him, or respect, or admiration, or esteem, although I think all those sentiments have been there in varying measures, but rather affection; and in saying that, which has been so very clear during the Symposium, and also at the Dinner, I think we would all like to include Genia Peierls in that affection. I think that it would be wholly inappropriate if I try to say anything else, because we are not here to "turn Rudi off" - we are here simply because a minor administrative phenomenon is going to take place at the end of the academic year, which will have no effect on Rudi's activities, I am sure, either professional or personal. We shall all continue to benefit for many years from the stimulation of both his academic company and his personal company. He is just going off, but he will be back again. So, may I ask you just to send him on his way, for the remainder of his scientific career, which I am sure will be a very long one, in the same way in which he began it - that is to say, with a tremendous Umklapp!!

D. H. Wilkinson

PEIERLS SYMPOSIUM
OXFORD 1974

1. Hone 2. Worden 3. Garcia 4. Ringwood 5. Barrett 6. Wong 7. Greenwood 8. Engel 9. Fall 10. Huber 11. Eggington

12. Williams 13. McCauley 14. Crowley 15. Cottingham 16. Kleppmann 17. Goncalves 18. Trueman 19. Byers

20. Thompson 21. Watson 22. Deans

23. Kocevar 24. Bauress 25. Golding 26. Pollard 27. Kobos 28. Mackintosh 29. Gehring 30. Brink 31. McMurry

32. Borysowicz 33. de Groot 34. Pethick 35. Korthals Altes 36. Norgett 37. M. J. Gillan 38. Field 39. Wheatley

40. Canto 41. Gomes 42. Lage 43. Brenner 44. Tsou 45. Chan

46. Donegan 47. Kirson 48. Leonardi 49. Stancu 50. Frampton 51. Millener 52. Krieger 53. Kurath 54. Talmi 55. Ng

56. Dodd 57. Epstein 58. Reading 59. O'Brien 60. Berman 61. D. Skyrme 62. Egri 63. T. H. R. Skyrme 64. Tait

65. Mary Gillan 66. Buck 67. Antoniou 68. Holcomb

69. Dunbar 70. M. Martinez 71. Filipe 72. Hasan 73. Gardiner 74. Corbett 75. Bhamathi 76. Choudhury 77. Buttery

78. Kirczenow 79. Ferreira 80. McDonald 81. Salpeter 82. Gunson 83. van Leuven 84. Lam 85. Copley 86. Ros

87. Harper 88. Schofield 89. Kearsley 90. Chretien

91. J. Martinez 92. Burkhardt 93. Zuker 94. Yoccoz 95. Belorizky 96. Curtis 97. Shepherd 98. Preist 99. Woodward

100. Matthews 101. Evans 102. Karl 103. Morrison 104. Butler 105. Kabir 106. Scharbach 107. Mrs. Post 108. Post

109. Fairbairn 110. Vinh-Mau 111. Tuttle

112. Smith 113. Wigginton 114. Castro 115. Jakeman 116. Major 117. Dowker 118. Gould 119. F. Lamb 120. D. Lamb

121. Norman 122. Kurti 123. Canning 124. Jones 125. Mihailovic 126. Grace 127. Hodgson 128. Bishop 129. Horgan

130. Grypeos 131. E. Sanderson 132. Stinchcombe 133. Law 134. Theumann

135. Tahir-Kheli 136. Paton 137. Aitchison 138. Brown 139. Mandelstam 140. Wills 141. Mondragon 142. Thouless

143. Edwards 144. Dalitz 145. Peierls 146. Frisch 147. Bethe 148. Ronald Peierls 149. Elliott 150. Wilkinson

151. Sciama 152. Radicati 153. McManus 154. Castillejo